ACTIVITY-BASED PHYSICAL SCIENCE

Kenneth W. Thompson, PhD

Emporia State University

KENDALL/HUNT PUBLISHING COMPANY
4050 Westmark Drive   Dubuque, Iowa 52002

Cover image courtesy Juice Drops.

CONTENTS

Part 1   Getting Started and Considerations for All Activity-Based Physical Science Units   1

1   A Preface Worth Reading   1

2   Safety   3

3   Concept Maps   4

4   Practice Concept Map   6

5   Long-Term Observations   7

6   Long-Term Observation of the Moon and the Nature of Science   9

7   Control Variable Experiments   12

8   Control Variable Experiment Format   14

9   Constellations, Stars, and Other Celestial Objects   15

Part 2   Introduction to Physical Science   18

10   Observations   18

11   Observations and Inferences   21

12   Black Box Activity   25

13   Making Measurements   26

14   Eye Chart   29

15   Astrolabe   31

16   Scientific Notation Practice   33

17   SI Conversions Practice   34

18   Graphs and Graphing   35

19   Calibrating a Balance   37

20	Accuracy, Reproducibility, Sensitivity, and Uncertainty	40
21	Meter Stick Balance	45
22	Make a Mobile	49
23	Introduction to Physical Science Questions and Problems	50
24	Introduction to Physical Science Reading	53
	Introduction	53
	Observations and Inferences	53
	Science and Mathematics	54
	Scientific Notation	54
	Measurement	56
	SI Units of Measurement	57
	Graphs and Graphing	59
	Locating Objects in the Sky	60
	Models	61
	Torque	61
	Describing Science	61
	Science and Society	63
	The Scientific Method	63
	Scientists and Science Careers – General	63
	Science Case Study – Harnessing Power	64
25	Introduction to Physical Science Concepts and Terms List	65
Part 3	More Physical Science	66
26	Comparing Total Mass	66

27 Comparing Masses and Mass of a Gas 68

28 Area and Volume 71

29 Densities of Solids 75

30 Densities of Liquids 78

31 Density of a Gas 81

32 A Real-World Example of Density: Wood 85

33 Sink, Float, Dive, Surface 89

34 Viscosity Control Variable Experiment 91

35 Strength and Elasticity Control Variable Experiments 94

36 Using Graphs 97

37 Mineral Identification 100

38 More Physical Science Questions and Problems 104

39 More Physical Science Reading 107

Matter 107

States or Phases of Matter 107

Properties of Matter 107

Density 108

Buoyancy 109

Changes in Matter 109

Law of Conservation of Mass 109

Significant Figures 110

Area and Volume 110

Circles 111

Slope   111

Minerals   111

Scientists   114

Science Careers   116

Science Case Study – Earth's Place in the Universe   116

40   More Physical Science Concepts and Terms List   118

Part 4   Chemistry   119

41   Solubility   119

42   Chromatography   122

43   Writing Formulas and Naming Compounds   124

44   Chemical Reactions   127

45   Introduction to Acids, Bases, and Indicators   132

46   Acid-Base Reactions – Practice 1   135

47   Acid-Base Reactions – Practice 2   137

48   Combining Acids and Bases   139

49   BTB Testing   140

50   Comparing the Acid in Two Types of Soda/Pop   142

51   Comparing the Neutralization Power of Antacids   145

52   Designing a Control Variable Experiment   148

53   Temperature, Heat Energy, and Evaporation   149

54   Phase Changes of Water   152

55   Ice Cream Chemistry   154

56	Water Quality Testing	156
57	Chemistry-Related Hobby Report	158
58	Scientists, Nature of Science, and History of Science Report	159
59	Solubility Questions and Problems	160
60	Thermal Energy Questions and Problems	162
61	Chemistry Reading	164
	Introduction	164
	Matter	164
	Atoms, Ions, Elements, and Compounds	164
	Elements and the Periodic Table	165
	Bonding	166
	Solutions	166
	Concentration	167
	Chromatography	167
	Chemical Reactions	167
	Acids, Bases, and Salts	168
	Acid Rain	169
	Indicators	170
	Water	170
	Water Pollution	170
	Chemistry and Society	172
	Heat and Temperature	172
	Kinetic Theory of Matter	173

	Phase Changes	173
	Scientists	174
	Science Careers	176
	Chemistry Case Study – Studying Fire and More	176
	Chemistry Case Study – Splitting the Atom and More	177
62	Chemistry Concepts and Terms List	178
Part 5	**Earth and Space Science**	**180**
63	Scale Model of the Solar System	180
64	Moon Phases	186
65	Time, Moon Phase, and Moon Location	190
66	Star Cycles Research Report	193
67	Planetarium and/or Observatory Visit	194
68	Simulating the Greenhouse Effect	195
69	Simulating Cloud Formation	198
70	Relative Humidity	200
71	Dew Point	203
72	Weather Map	205
73	Relative Time	207
74	Mineral Shapes	211
75	Rock Identification	214
76	Soil	219
77	Studying Fossils at a Geology Museum	221
78	Plate Tectonics Illustrations and Explanations	228

79	Geoscience Maps	232
80	Investigating Earth Science Hazards	234
81	Earth and Space Science-Related Hobby Report	235
82	Scientists, Nature of Science, and History of Science Report	236
83	Earth and Space Science Questions and Problems	237
84	Earth and Space Science Reading	239
	Introduction	239
	Astronomy	239
	Geology	245
	Hydrology	249
	Meteorology	250
	Scientists	255
	Science Careers	256
	Earth and Space Science Case Study – Plate Tectonics	256
	Earth and Space Science Case Study – Extending Time	256
85	Earth and Space Science Concepts and Terms List	257
Part 6	Physics	259
86	Work and Power	259
87	Speed, Velocity, Acceleration, Momentum, Potential and Kinetic Energy	262
88	Bicycles: Studying a Compound Machine	266
89	Motion Questions and Problems	270
90	Relative Motion Questions and Problems	272

91	Light a Bulb	273
92	Conductors and Insulators	276
93	Series and Parallel Circuits	278
94	Magnetism	280
95	Cooling	283
96	Heat and Insulation	287
97	Explaining Sunset Colors	289
98	Polaroid Filters	291
99	Reflection and Refraction	293
100	Distance and Light Intensity	295
101	Orientation and Energy per Unit Area	297
102	Sound	299
103	Half-life Simulation	301
104	Calculating Radiation Dosages	304
105	Physics-Related Hobby Report	306
106	Scientists, Nature of Science, and History of Science Report	307
107	Physics Questions and Problems	308
108	Physics Reading	310
	Force	310
	Work	310
	Power	310
	Machines	310
	Motion	311

Energy   312

Kinetic and Potential Energy   313

Heat   314

Electricity   314

Magnetism   315

Light   315

Sound   316

Nuclear Energy   316

Scientists   317

Science Careers   317

Physics Case Study – Laws for Heaven and Earth   318

Physics Case Study – Matter, Energy, Time, and Space   318

109   Physics Concepts and Terms List   319

110   Activity-Based Physical Science References and Additional Reading   320

Part 1   **Getting Started and Considerations for All Activity-Based Physical Science Units**

1   **A Preface Worth Reading**

To the Student:

This textbook is designed for you as you study and experience the physical sciences. The main goal of *Activity-Based Physical Science* is to provide opportunities for you to develop understanding of basic chemistry, earth and space science, and physics concepts. Activities have been selected to address applicable science content standards as identified by the National Research Council's (1996) *National Science Education Standards*. All students must take advantage of the opportunities presented to them to learn and understand subject matter disciplines that make up the physical sciences. Although laboratory activities will be a featured mode of learning in this textbook, you will be involved in a variety of modes of learning science as outlined by Farmer, Farrell, and Lehman (1991). In addition to laboratory activities, it is expected that other modes of learning science will include: demonstrations, discussions, lectures, questions and answers, audio-visual and technological activities, group projects, individual projects, supervised practice, and homework.

Science Understanding

Science understanding can be described in a variety of ways. This textbook will provide opportunities for students to achieve different levels of science understanding as described by the College Entrance Examination Board (1986). The College Entrance Examination Board outlines four hierarchical levels of science understanding. The first level of understanding, called the **descriptive** level, involves observation, i. e., using the senses or extensions of the senses. Using a telescope or describing what one might see, smell, or hear as a chemical reaction takes place are examples of the descriptive level of science understanding. The second level of science understanding, the **qualitative** level, extends observation by trying to provide explanations for the phenomena observed. The **quantitative** level of science understanding builds on the descriptive and qualitative levels by making use of numbers and measurements. The fourth level of science understanding is the most abstract. The **symbolic** level of science understanding builds on the three other levels by making use of formulas, symbols, and equations. Many scientists have a goal of trying to achieve science understanding at the symbolic level. Why do you think scientists might want to use formulas or equations to describe physical science phenomena? Success for students at this level of science understanding builds on earlier levels and should mean more than simply putting numbers into a formula to derive an answer. Many of the laboratory activities presented in this text require students to observe, explain, and use mathematics. Do think about the four levels of science understanding as you complete activities assigned to you by your instructor.

Format

As stated earlier, a goal of this text is to provide opportunities to develop science understanding and not memorization of many seemingly unrelated facts. Principally, this will be accomplished by the completion of laboratory activities. As much as possible, laboratory activities will precede discussion of the physical science concepts to be understood and they will take the form of a learning cycle or conceptual change format. Abraham (1992) and Kyle, Abell, and Shymansky (1992) each build on the work of others and suggested formats for learning and understanding science that are useful in an activity-based approach. Science programs that have an activity oriented format are sometimes referred to as "hands-on, minds-on, hearts-on" programs to take into account the knowledge (cognitive), psychomotor (physical skills associated with knowledge), and affective (values, attitudes) learning domains. In a conceptual change approach to learning science suggested by Kyle, Abell, and Shymansky (1992), students are first asked to identify what they know about a particular science concept. The laboratory activity that follows provides context as well as serving as a motivational and problematic experience related to the science concept under discussion. Once the laboratory activity is completed, students exchange views and compare ideas. A part of this step of learning and developing science understanding includes students citing evidence for their views. Finally, students are asked to use the scientific conceptions by applying them in different settings. Abraham (1992) described a **learning cycle** approach to the learning and teaching of science. Robert Karplus and his colleagues are given credit as the originators of the learning cycle during the development of the Science Curriculum Improvement Study (SCIS). In the **exploration** phase, students explore different physical science phenomena usually by doing a laboratory activity as an introduction to the concept. In the **concept development** phase, students share their ideas in discussion with fellow students in cooperative learning groups and their instructor in a discussion format to identify the concepts and to develop scientifically acceptable views of the concepts. As the name implies, the **application** phase of the learning cycle requires students to apply the concepts to demonstrate their understanding. Applying the concept may involve more laboratory work, solving problems, reading, or extending the concept into different situations.

In this textbook, students have the opportunity to work on and complete laboratory activities that may be classified in three ways: 1) illustrations or **demonstrations** of known physical science phenomena; 2) **experiments** where students **control variables**; and, 3) **long term observations** where students make a series of observations and then try to make sense of the observations. Students should approach each activity as a way to enhance their understanding of physical science concepts and of physical science.

Typically, students will work on the laboratory activities in **cooperative groups** in ways described by Johnson, Johnson, Holubec, and Roy (1986); this means more than simply working with a lab partner. Five basic elements must be in place for the groups to be considered cooperative. Among group members there will be **positive interdependence**. This interdependence manifests itself in shared resources and materials, shared division of labor with each group member contributing to a final product, and no one group member completing more than their share of the responsibilities. **Face-to-face interaction** is a basic element of cooperative learning.

Group members will interact with one another, share their experiences, and share their understanding of science concepts. **Individual accountability** is a cornerstone of cooperative learning. In some ways, cooperative learning is a bit of a misnomer. Although experiences and meaning can be shared, learning does depend on the individual. It is the responsibility of the individual to learn and understand the material. A fourth basic element is **interpersonal and small group skills**. It is assumed that students possess the social skills for collaboration. Often when a group does not function well, it is because individuals in the group fail to live up to their individual responsibilities. Finally, **processing** or analyzing how well individuals and groups are functioning is the last of the basic elements of cooperative learning. Effective working relationships must be achieved and maintained.

To quickly summarize, laboratory activities completed in a learning cycle format while working cooperatively are key features of the format of this course. Students can expect to spend about two-thirds of their time in class working on laboratory activities and about one-third of their time in lecture, questions and answers, or discussion formats.

Student Success

The author of this text feels that students that conform to the "6Ps" regarding personal behavior that follow are most likely to achieve success in this course. The "6Ps" include: 1) be prompt – show up for class on time and turn in assignments on time; 2) be prepared – have all materials and be mentally and physically ready to work; 3) be productive – showing up on time is not enough, students must complete assignments and demonstrate competence; 4) be persistent – learning is not always easy and fun, do not give up, some things require hard work; 5) be positive – the world already has enough people that are negative and can only find fault; and, 6) be polite – you, your fellow students, and your instructor are all worthy of respect.

There are at least four things that students can do to help them achieve success in this course. "Time-on-task" or spending quality time both doing and thinking about physical science in and outside of class is probably the most important. Second, students need to identify occasions where they need help and then take the initiative to get help when it is required to complete tasks. Ask questions if you do not understand how to complete tasks. People are different and not all topics are equally interesting to all students. Given this, students must try to understand why some people view certain tasks as worthwhile. To achieve success, you must occasionally experience some success. Certainly, there are other behaviors that assist one to be successful.

2   Safety

Safety in the science classroom is a priority for students, teachers, and administrators. Your teacher should provide proper instruction regarding safety, proper supervision during science activities, and your instructor should properly maintain equipment used. **When conducting activities, teachers and students must follow proper safety precautions! Students and teachers should make safe choices and use procedures that meet local, state, and federal safety guidelines.** Do not assume all safety warnings and precautions are included.

Concept Maps

Concept maps have a variety of uses in science education. One way concept maps will be used in this course is to assess and evaluate student understanding of physical science concepts. Concept maps illustrate key concepts organized hierarchically from most general to most specific with linking words that relate concepts to each other in a particular context.

Novak and Gowin (1984) highlight six important features of a concept map. A sample concept map constructed by a student is illustrated in Figure 1-1. Figure 1-1 contains one or more examples of each of the six important features.

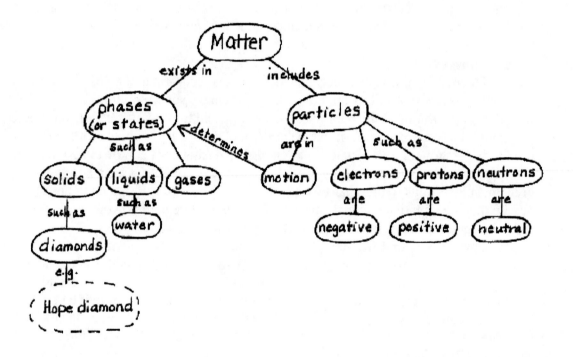

Figure 1-1. Sample concept map

A simple definition for a **concept** is: an object or an event that represents a generalization. On the sample concept map, concepts are enclosed by solid ellipses. By this definition, "dog" and "running" are each considered concepts. Do you understand why? Also, by this definition, the Golden Gate Bridge would not be considered a concept. Do you understand why not?

A concept map is hierarchical. Concepts are listed from **most general** (at the top of the page) to **most specific** (at the **bottom** of the page). Concepts at the same level represent nearly the same level of generality or specificity.

An **example** is a specific object or event, i. e., an object or event that does not represent a generalization. Examples are enclosed by dashed ellipses on the sample concept map. It should be clear to you that "diamond" represents a concept and the "Hope Diamond" represents an example.

Linking words are words that illustrate the relationship between one concept and another concept. Linking words are very important parts of a concept map. Without linking words, a concept map is nothing more than a bunch of terms that may or may not be organized in a way that shows understanding. Just looking at the top left side of the sample concept map, "exists in" and "such as" are examples of linking words that illustrate the relationship between "matter" and "phases (or states)" and "solids" and "phases (or states)," respectively.

Cross-linking words are linking words that "cut across" a concept map. In Figure 1-1, the segment or theme on the left side of the concept map is phases or states of matter. The right side segment or theme is particles of matter. "Determines" is a cross-linking word that illustrates the relationship that the motion of particles determines whether the matter exists as a solid, liquid, or gas.

As you move down a concept map, a concept linked to another less general concept represents a **level**. A concept connected to an example does not constitute a level.

As was stated earlier, one way concept maps will be used in this course is as a measure of student understanding. On a relative point scale, Novak and Gowin (1984) described a scoring system where one point is awarded for each valid concept linked to another concept. Also, one point is awarded for each valid example for a concept, but this is limited to one example per concept. Recall that concept maps are supposed to show understanding. Listing several examples of a concept is not considered to be demonstrating greater conceptual understanding than being able to list one. Five points are awarded for each valid level, i. e., each subordinate concept is more specific and less general than the concept drawn above. Three levels are represented in Figure 1-1. Can you identify them? Examples do not count as levels! Ten points are awarded for each cross link that is valid and considered significant and two points are awarded for each cross link that is valid but does not illustrate a synthesis between sets of related concepts.

Concept maps have other uses in science education besides as assessment devices. Some students may find it helpful to use concept maps as an organizing tool that helps them see the "big picture" or to plan experiments. Some may use concept maps as a study strategy. Students should not memorize concept maps!

4   **Practice Concept Map**

      Select a general topic about which you consider yourself knowledgeable. The topic need not be a science topic. Identify more specific concepts associated with the general topic. For practice, construct a concept map that has at least 3 levels. Use the criteria listed below to evaluate your concept map.

General to Specific	OK	Not OK	Not Use	Comments:
Concepts	OK	Not OK	Not Use	Comments:
Linking Words	OK	Not OK	Not Use	Comments:
Examples	OK	Not OK	Not Use	Comments:
Levels	OK	Not OK	Not Use	Comments:
Cross Linking Words	OK	Not OK	Not Use	Comments:
Clarity	OK	Not OK	Not Use	Comments:

      Because people are different, some need more practice than others to become confident in their ability to construct concept maps. Perhaps your instructor will ask that you construct other practice concept maps or you may need to practice on your own.

5   Long-Term Observations

Long-term observations require students to make and record observations, analyze data, and then make inferences or conclusions about the data. In nearly all instances, observations will be made outside "normal" course meeting times and will extend beyond the typical time requirements for in-class science activities. Students must confirm their plans for completing long-term observations with their instructor prior to beginning any long-term observation activities.

Long-Term Observations

Related National Science Education Standards will depend on options selected but minimally, they will include:

Teaching Standard A:	inquiry-based
Science Content:	systems, order, and organization
	evidence, models, and explanation
	constancy, change, and measurement
	evolution and equilibrium
	form and function
Science as Inquiry:	abilities to do scientific inquiry
	understanding about scientific inquiry
History and Nature of Science:	science as a human endeavor
	nature of science
	nature of scientific knowledge
	historical perspectives

Background Information:

One way that science operates is by one or more scientists collecting observations over an extended period of time. Scientists try to make sense out of observations by analyzing data and determining if there are patterns of change. An important part of the process of science is communicating with other scientists.

As part of the requirements for this course you will need to make observations over an extended period of time where physical science variables (e. g., temperature, light, . . .) play a significant role in the investigation. Included here are some options and your instructor can help suggest other possibilities. After you have made the required number of observations as stated by your instructor, you need to make some inferences based on your observations, and turn in a written report. Your inferences or interpretations are attempts to identify and describe relationships, if any, among the variables you have observed. Again, your instructor will provide details about the inferences for the various long-term possibilities. What follows is a non-inclusive list of long-term observation possibilities.

1. Moon phase, Moon location, and time of day – This long-term observation is required of all students. Instructions are provided on a separate page.

2. Weather – Numerous variables may be investigated in a weather long-term observation. Plot the data for a seven day period on one graph. The x-axis or abscissa will indicate day and time by the hour. Use a scale for the y-axis or

ordinate that allows you to plot: temperature, wind chill or heat index, dew point, relative humidity, and wind speed using the same scale. Use a different scale (an expanded scale with a low of 27 inches of mercury and a high of 33 inches of mercury) to indicate pressure. Also, below the x-axis, indicate wind direction and precipitation amounts when available. After graphing the data, you need to describe the relationships that exist, if any, between various pairs of weather variables. Weather data need to be collected over seven consecutive days.

3. The positions of stars and time will require nighttime observing. This long-term observation can be completed in two ways. One possibility is to observe and record the positions of stars in a selected area of the sky. The other possibility is to observe and record the positions of stars in a few constellations. In either case, you need to note positions of the stars, as well as observation dates and times on 14 days spread out over a period of about two months. You can use an astrolabe to help note positions of the stars. Make and record observations about every four days. After analyzing the data, you need to identify the relationships, if any, between stars' positions and time.

4. The relationship of light and temperature during daylight hours requires use of some special equipment. This long-term observation will require the use of a graphing calculator and a CBL (Calculator Based Laboratory). Using equipment and a program specified by your instructor, you will collect temperature and daylight intensity data on two separate days. After analyzing the graphed data, you need to identify the relationship, if any, between temperature and light intensity.

5. Sunset times and positions the Sun sets (or Sunrise times and positions the Sun rises) – CAUTION: DO NOT LOOK DIRECTLY AT THE SUN! Check with your instructor before selecting this option. This long-term observation requires you to keep track of sunset or sunrise positions and the time of sunset or sunrise. Before attempting to work on this activity, check with your instructor to make sure that you know and can practice proper procedures and will complete the activity following safe practices. To record the position of sunset, you first need to sketch your view of your western horizon (eastern horizon for sunrise). You need to note the position of the Sun at sunset (or sunrise) and the time. Do this for 14 days spread out over a period of about two months. This means you will be making and recording observations about every four days. After analyzing the data, you need to identify the relationships, if any, between sunset (or sunrise) positions and times.

6. Your instructor may suggest other possibilities or you may think of and have ideas of your own. Make sure your instructor approves your idea before you start.

Related National Science Education Standards:
Teaching Standard A: inquiry-based

Science Content:	systems, order, and organization
	evidence, models, and explanation
	constancy, change, and measurement
	evolution and equilibrium
	form and function
Science as Inquiry:	abilities to do scientific inquiry
	understanding about scientific inquiry
Physical Science:	properties of objects and materials
	position and motion of objects
	motions and forces
	light, . . .
Earth and Space Science:	objects in the sky
	changes in Earth and sky
	structure of the Earth system
	energy in the Earth system
Science in Personal and Social Perspectives:	changes in environments
History and Nature of Science:	science as a human endeavor
	nature of science
	nature of scientific knowledge
	historical perspectives

1. Use the Moon record sheet to carefully record the date and time and to sketch the appearance and location of the Moon as you face south. If the moon lies on an imaginary east-west line through the overhead point, sketch the Moon on the arc that makes the semi-circle. If it lies south of the east-west line through the overhead point, sketch it in its approximate position inside the semi-circle. If the Moon lies north of the east-west line through the overhead point, sketch it in its approximate position outside the semi-circle.

2. Each time you observe the Moon make additional notes to thoroughly describe what you observe.

3. If you work as an individual, you will need to make a total of 20 observations of the Moon. If you work in a cooperative group of up to 4 persons, you will need to make a total of 32 observations. It is OK, in fact you are encouraged, to occasionally make two observations of the Moon on the same date separated by a "reasonable" amount of time. For the second observation to count, you must observe at least one difference in the Moon from the first observation.

4. After observations are completed, you need to type/word process a short paper that *summarizes your observations* and *includes ten inferences* made from your observations. If outside sources are used, you must properly reference those sources. Also, you need to include a detailed sketch or picture of the full Moon.

5. The tentative due date for completion of the above is by the end of the 11th week of classes or at a time designated by your instructor.

Record Sheet for Moon Observations

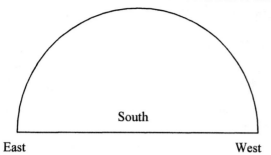

Date: Time:

Observation Notes:

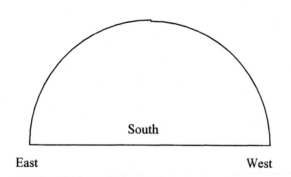

Date: Time:

Observation Notes:

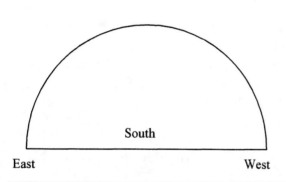

Date: Time:

Observations Notes:

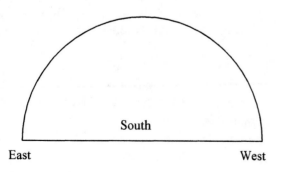

Date: Time:

Observation Notes:

Record Sheet for Moon Observations

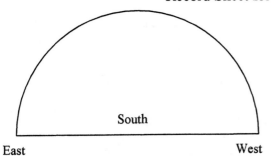

South

East   West

Date: Time:

Observation Notes:

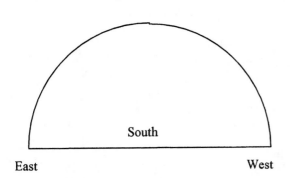

South

East   West

Date: Time:

Observation Notes:

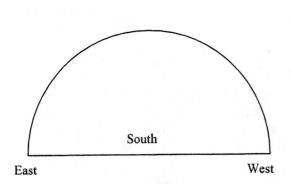

South

East   West

Date: Time:

Observations Notes:

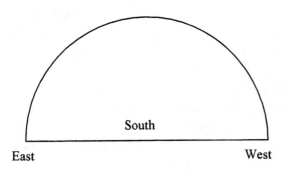

South

East   West

Date: Time:

Observation Notes:

Students are required to design and complete control variable experiments. In some instances, planning and carrying out control variable experiments may extend beyond "normal" course meeting times. Background information follows and the specific write-up format is provided in activity #8 "Control Variable Experiment Format." Students must confirm their plans for completing control variable experiments with their instructor prior to beginning any control variable experiment activities.

Control Variable Experiments

Related National Science Education Standards will depend on options selected but minimally, they will include:	
Teaching Standard A: inquiry-based	
Science Content:	systems, order, and organization
	evidence, models, and explanation
	constancy, change, and measurement
	evolution and equilibrium
	form and function
Science as Inquiry:	abilities to do scientific inquiry
	understanding about scientific inquiry
History and Nature of Science:	science as a human endeavor
	nature of science
	nature of scientific knowledge
	historical perspectives

Background Information

Individual words can have different meanings to different people. In this course a distinction will be made between an "experiment" and a "demonstration." Generally, think of an experiment as a laboratory activity where the outcome is unknown. Think of a demonstration as showing or illustrating a known scientific idea. It is possible that an experiment to one person might be a demonstration to another person. Would you classify most of the laboratory activities that you have done as experiments or demonstrations?

Scientists conduct control variable experiments as a way to determine cause and effect. They do this most simply by keeping all variables or factors the same except for one. The familiar "scientific method" that a number of science texts make reference to can most closely be associated with conducting control variable experiments. A scientist may have done some observing and thinking about observations made. This could lead to a problem statement or asking a question. Asking significant questions is an important aspect of creativity in science. When a question is asked or a problem stated, the person posing the question or stating the problem often has some idea of the potential relationship among the variables of interest. A hypothesis is an educated guess or prediction of the relationship between two variables of interest. A test or experiment is designed to see what relationship, if any, exists. The conclusion may be that the hypothesis is valid. Or, it could be that the hypothesis needs to be discarded, i. e., it was invalid or it may be the hypothesis needs to be changed. Depending upon the conclusion, the knowledge gained may be put to some practical purpose.

Asking significant questions is an important starting point for conducting a control variable experiment. Some, but not all, possible questions or problem statements include:

What is the effect of _____ upon _____ ?

What evidence exists that _____ ?

After a scientific observation of _____ it is evident that _____ .

As part of this course, you will need to complete control variable experiments where at least one of the important variables is a physical science variable. Use the form that follows to brainstorm control variable experiment ideas and to determine which variables will be unchanged and which variable will be changed.

Control Variable Experiment Requirements

Your write-up of a control variable experiment must include:
1. the question being investigated or a statement of the problem.
2. your hypothesis.
3. a brief description of your procedure.
4. communication of your results. Often this involves making graphs.
5. your conclusion.

Your instructor can tell you how many control variable experiments need to be done. Discuss your ideas with your instructor and get permission from your instructor before you proceed. *Remember to consult with your instructor before starting any control variable experiment.*

Control Variable Experiment Ideas

Problem Statement or Question	Variable(s) that will stay the same	Variable(s) that will be allowed to change

8   Control Variable Experiment Format

Use the format listed here to complete your write-up of one or more control variable experiments.

This control variable experiment (select one):

_____investigates a problem where the conclusion is unknown to me.

_____illustrates a known scientific principle.

_____is appropriate for the _____ level.

• The question being investigated or a statement of the problem:

• Your hypothesis:

• Brief description of your procedure:

• Communication of your results (often this involves making graphs):

• Your conclusion:

Related National Science Education Standards:	
Science Content:	systems, order, and organization
	evidence, models, and explanation
	constancy, change, and measurement
	evolution and equilibrium
Science as Inquiry:	abilities to do scientific inquiry
	understanding about scientific inquiry
Physical Science:	position and motion of objects
	motions and forces
	light, . . .
Earth and Space Science:	objects in the sky
	changes in Earth and sky
	structure of the Earth system
	energy in the Earth system
Science in Personal and Social Perspectives:	changes in environments
History and Nature of Science:	science as a human endeavor
	nature of science
	nature of scientific knowledge
	historical perspectives

Background Information

Other than as a convenience for naming parts of the sky, parts of a sky map, or sections of a star chart, constellations have no particular significance to modern astronomy. Stars appear to remain in fixed positions relative to one another and revolve around a fixed point in the sky marked by the Pole Star or North Star (Polaris). This led different ancient cultures to identify groups of stars as constellations. Some ancient astronomers or religious leaders associated constellations with animals or gods. Stars in a constellation may appear close together in the night sky, but in actuality may vary a great deal in how far they are from Earth. Many stars in constellations move independently from one another but there are stars affected by the gravity of other stars. Over geologic time these motions are significant and would change the shape of constellations but would not be significant over a human lifetime.

Most constellations appear during some seasons and disappear during other seasons. Circumpolar constellations are exceptions. Circumpolar constellations exist separately for each of the Northern and Southern Hemispheres. During unobstructed viewing, circumpolar constellations can be seen at night all year long. Along with cycles of the Sun and Moon, ancient astronomers used the cyclical nature of motion of the stars to mark the seasons.

What follows is a list of some of the circumpolar constellations and seasonal constellations. Do not interpret the constellations listed under a season as being only visible during that season. They are listed when they are most prominent in the night sky and can be seen during more than one season but not throughout the year. Your instructor can provide star charts that help identify outlines of the constellations and their relative positions. Of course, you are strongly encouraged to try to identify constellations in the night sky! In addition, your instructor will note any individual stars or other astronomical phenomena you will need to identify.

Circumpolar Constellations	Ursa Major (double stars: Mizar, Alcor; pointer stars: Dubhe, Merak)
	Ursa Minor (pole star: Polaris)
	Cassiopeia (Caph)
	Cepheus
	Draco (Thuban)

Spring Constellations	Canis Major (Sirius)
	Canis Minor (Procyon)
	Gemini (Castor, Pollux)
	Cancer (Beehive Cluster)
	Leo (Denebola, Regulus)

Summer Constellations	Bootes (Arcturus)
	Libra
	Scorpius (Antares)
	Sagittarius
	Corona Borealis
	Hercules (M13 Cluster)

Fall Constellations	Lyra (Vega)
	Cygnus (Deneb)
	Aquila (Altair)
	Capricornus
	Aquarius
	Pegasus

Winter Constellations	Andromeda
	Aries
	Pisces
	Perseus (clusters)
	Auriga (Capella)
	Taurus (Aldebaran, Pleiades Cluster)
	Orion

Astronomy and astrology are NOT the same! Astronomy is a science and astrology is a pseudoscience. Do not confuse the two! Often students are interested in the constellations of the Zodiac. For your information a list is provided.

Zodiacal Constellations

Aries – the Ram
Taurus – the Bull
Gemini – the Twins
Cancer – the Crab
Leo – the Lion
Virgo – the Virgin
Libra – the Scales or Balance
Scorpius – the Scorpion
Sagittarius – the Archer
Capricornus – the Goat
Aquarius – the Water Bearer
Pisces – the Fish

Name(s)   _____

_____

_____

_____

10   Observations

Related National Science Education Standards:
Teaching Standard A: inquiry-based
Science Content: systems, order, and organization
evidence, models, and explanation
constancy, change, and measurement
evolution and equilibrium
form and function
Science as Inquiry: abilities to do scientific inquiry
understanding about scientific inquiry

Exploration/Awareness:

Observing involves **using the senses**, i. e., sight, smell, touch, taste, and hearing, or an extension of the senses. Examples of extending the sense of sight include using a telescope to see distant objects or using a microscope to see very small objects. Sometimes observations are referred to as facts. Does everyone see or hear at the same level? By contrast an **inference** is an interpretation or **conclusion based upon observation**.

In this activity you will be asked to make a number of observations. Your instructor will identify which observation activities you need to complete.

1. Observe the words on the index card. Note that the large, stoppered test tube is filled with a colorless, clear liquid. Observe the words on the index card while looking through the large, liquid-filled, stoppered test tube. Record what you observe in the space below.

2. Your instructor will alternately hold two pieces of plastic for you to observe. Note your observations in the space that follows.

[ ]

3. Your instructor will show you a scientific apparatus and use it to demonstrate a known scientific principle. Note your observations in the space below.

[ ]

4. Your instructor will provide some "pellets" suitable for tasting by most persons. If you are unable to taste the "pellets," observations may be shared by one of your lab partners. Note your observations in the space below.

[ ]

Concept Development:
5. Does observing mean the same thing as seeing? Explain.

6. What physical science phenomena are involved in what you observed in observations 1, 2, 3, and 4?

Application:
7. What sense do most humans rely on most of the time?

8. Identify an organism (other than humans) that relies on a well developed sense of:
    a. sight

    b. smell

    c. touch

    d. taste

    e. hearing

9. How might a heavy reliance on the sense of taste be precarious?

10. Briefly describe an observation activity that would require the use of a sense other than sight.

11. Very carefully observe an object or event that you previously had not observed closely. What did you notice that was unexpected?

12. Describe or illustrate a situation where one or more of your senses was fooled or "tricked."

Background Information:
   For additional background, read the **Observations and Inferences** section of the Introduction to Physical Science Reading.

11   Observations and Inferences

Related National Science Education Standards:
Teaching Standard A: inquiry-based
Science Content: systems, order, and organization
evidence, models, and explanation
constancy, change, and measurement
Science as Inquiry: abilities to do scientific inquiry
understanding about scientific inquiry

Exploration/Awareness:

After reading the preface, you know that observation is a beginning point for developing science understanding. However, the importance of observing is not limited to science. Can you imagine situations where careful observation is important? Are observations facts?

The activity that follows is based on a classic activity using tracks that shows up in various publications. The author's review found the earliest copyrighted appearance of the activity in:

American Geological Institute. (1965). Investigating the earth (laboratory manual pp.19-4 – 19-5 and teacher's guide pp. 19-21 – 19-24). Denver, CO: Smith-Brooks Printing Company (for laboratory manual) and Boulder, CO: Johnson Publishing Company (for teacher's guide).

Use a different tracks diagram, Figure 2-1, to determine whether the following statements are observations or inferences and circle the appropriate letter. Assume: 1) the tracks were made in a northern state in the United States during January, 2) the tracks were made by a living organism, and 3) the tracks were made by two different organisms, i. e, they were not made by the same individual. Note that horizontal lines are used to divide the total area into three numbered areas and north is at the top of the page.

O   I   1. There are two sets of tracks in area 2.

O   I   2. One set of tracks was made by a skier.

O   I   3. One organism was in areas 1, 2, and 3.

O   I   4. The organisms are the same size.

O   I   5. The organisms were both in area 2 at the same time.

O   I   6. One organism ran into the other organism.

O I   7. One organism traveled from west to east.

O I   8. One organism traveled from north to south.

O I   9. The northern area is at a higher elevation than the southern area because the skier is going downhill from north to south.

O I   10. The tracks were made in snow.

O I   11. The organism with "straight line" tracks was in area 2 after the other organism.

O I   12. Because the tracks are different, the organisms are not closely related.

Concept Development:

13. What distinguishes an observation from an inference?

14. How can you tell from footprints whether a person walked forwards or backwards? Walk backwards and forwards and give some careful thought to any differences that might show up in the footprints.

15. How does the fact that people typically wear shoes affect inferences that can be made from human "footprints?"

Application:

16. Assume the average person walks 6 miles per day for a life span of 80 years and that the average step is three feet long. If only one footprint in a billion is fossilized, how many fossil footprints would the average person produce? Show your work! Additional space is provided on the next page.

17. What does "don't judge a book by its cover" mean?

18. During a legal trial, for something to be considered a fact and used as evidence, must it be observed? Explain and give an example to support your explanation.

Background Information:

   For additional background, read the **Observations and Inferences** section of the Introduction to Physical Science Reading.

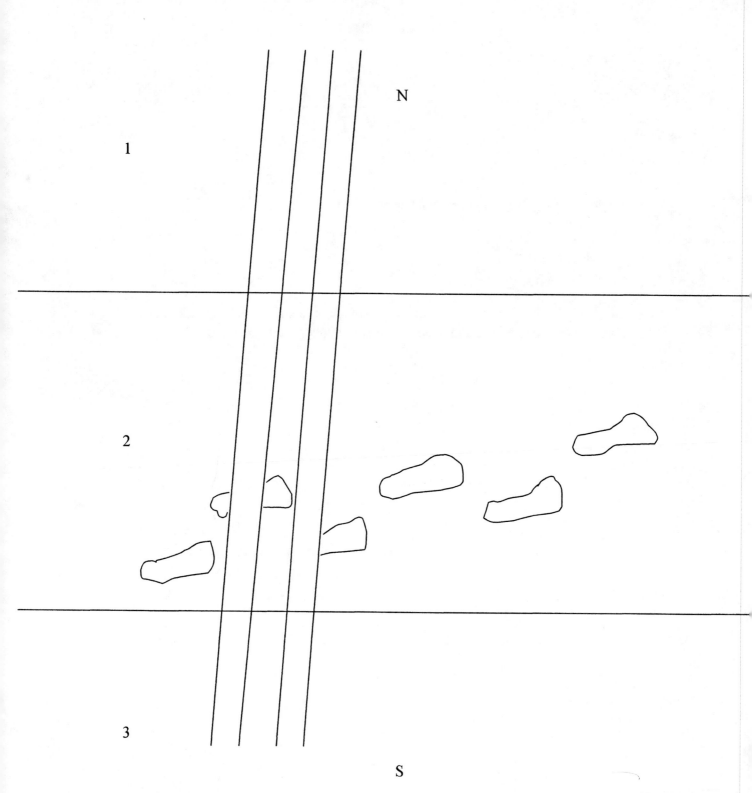

Figure 2-1. Tracks

12   Black Box Activity

Related National Science Education Standards:
 Teaching Standard A: inquiry-based
 Science Content: systems, order, and organization
                                        evidence, models, and explanation
                                        constancy, change, and measurement
 Science as Inquiry: abilities to do scientific inquiry
                                        understanding about scientific inquiry

Exploration/Awareness:

Making observations is an important part of doing science. Scientists can make direct observations of some phenomena. Other phenomena may require the use of technology to extend the capabilities of human senses or to make indirect measurements.

1. Use the black box and probes provided, to determine what the inside of the black box is like without damaging the inside or the probes.

Concept Development:

2. Use easy-to-obtain materials to design and build a three dimensional model of the inside of the black box.

Application:

3. How is your investigation of the black box activity analogous to scientists investigating:

   a. the structure of an atom?

   b. the surface of a distant planet?

   c. the ocean floor?

Background Information:

For additional background, read the **Observation and Inferences** and **Models** sections of the Introduction to Physical Science Reading.

13   Making Measurements

Related National Science Education Standards:
Teaching Standard A: inquiry-based

Science Content:	systems, order, and organization
	evidence, models, and explanation
	constancy, change, and measurement
	evolution and equilibrium
	form and function
Science as Inquiry:	abilities to do scientific inquiry
	understanding about scientific inquiry
Physical Science:	properties of objects and materials
Science and Technology:	understanding about science and technology

Exploration/Awareness:

Measuring is something that scientists do, but more importantly, it is something that affects everyone. Have you been to the grocery store lately? Have you purchased gasoline for your automobile? Have you put air in your bicycle tires lately? It is not only scientists that measure, people measure! What are some things that people measure? What do we measure these things with?

Prior to starting this activity, your teacher will give you some brief instructions on how to properly record measurements. To represent a measurement, the convention that is required in this course is to record a **number** and a **unit**. When recording a measurement, record the digits that you observe plus one additional digit that you estimate.

In this activity, you are asked to measure the objects made available to you by your instructor using the measuring devices provided. Properly record your measurements in the table that follows. The measuring devices provided include: centimeter ruler, meter stick, triple beam balance, graduated cylinder, and syringe.

Table 2-1   Making Measurements

Object	Device used to make the measurement	Measurement
vitamin C tablet		
string		

Object	Device used to make the measurement	Measurement
width of room		
liquid in a soda/pop can		
piece of paper		
liquid in a cup		
aspirin tablet		

Concept Development:

1. Identify the general steps that you went through to make a measurement and compare your thoughts to those listed in the background reading on measurement.

2. Check with other groups that measured at least some of the same objects that you measured. Are measurements of the same object exactly alike or different?

3. If some of the measurements of the same objects are different, how different are they?

4. When measuring, identify some potential sources of error.

Application:

5. Identify and list a measured attribute that affects you on a regular basis. What measuring device is used to measure this attribute?

6. Illustrated below is a segment of a centimeter ruler. The length of an object is marked by the arrow. Properly record the measurement.

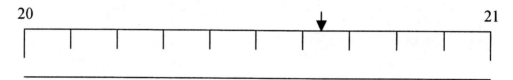

Measurement recorded as: _____

7. Illustrated below is a segment of a graduated cylinder. The volume of a liquid is marked by the arrow. Properly record the measurement.

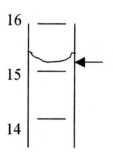

Measurement recorded as: _____

Background Information:

For additional background, read all sections of the Introduction to Physical Science Reading related to measurement.

Name(s)

_____

_____

_____

_____

14   Eye Chart

Related National Science Education Standards:

Teaching Standard A: inquiry-based	
Science Content:	systems, order, and organization
	evidence, models, and explanation
	constancy, change, and measurement
	evolution and equilibrium
	form and function
Science as Inquiry:	abilities to do scientific inquiry
	understanding about scientific inquiry
Physical Science:	properties of objects and materials
	light, . . .
	interactions of energy and matter
Life Science:	structure and function in living systems
Earth and Space Science:	objects in the sky
Science and Technology:	understanding about science and technology
Science in Personal and Social Perspectives:	personal health

Exploration/Awareness:

Can numbers be associated with observing? It is hard for most of us to associate numbers with the senses of touch, taste, and smell. However, we can imagine numbers being associated with vision, e. g., 20/20 vision, and with hearing, e. g., decibel levels associated with sound. In this activity you will have the opportunity to take a closer look at how eye charts are constructed. Optometrists use eye charts as an aid to determine acuity, or the sharpness of vision. We have indicated that sight is the dominant sense for most persons. Do all persons observe equally well? Are observations facts?

A Snellen eye chart has letters on it that decrease in size as you move from the top of the chart to the bottom of the chart. Along the right side of some charts are numbers that correspond to the distance that a person with "normal" vision could stand and still see that size of letter. The chart is usually read at a distance of 20 feet. Acuity is represented as a fraction with the top number (numerator) the distance at which the viewer stands and the normal maximum legible viewing distance as the bottom number (denominator) of the fraction. Thus, if at 20 feet you can read letters on the row marked "40," your visual acuity is 20/40 or ½ normal. If, standing at 15 feet, you can read letters on the row marked "40," your visual acuity would be 15/40 or 3/8 normal.

The International Society for the Enhancement of Eyesight (ISEE) maintains a website (available January 4, 2002) at <http://www.i-see.org/> that provides vision information and gives specifications given for an eye chart available at a <http://www.i-see.org/eyecharts.html>. Those specifications follow:

Distance (feet)	70	60	50	40	30	20	15	10	7	4
Letter height (mm)	31	27	22	18	13	9	7	4	3	2

1. Construct a graph using distance and letter height data.

Concept Development:
2. Should the graph represent a straight line? Why or why not?

3. How might acuity affect visual observations?

4. How do you suppose "normal" maximum legible viewing distance was established?

5. Establishing "normal" vision parameters would correspond to what part of a control variable experiment?

Application:
6. Other than this activity, identify a situation where acuity would be important.

7. Interpret 20/40 vision and 40/20 vision.

Background Information:
   For additional background, read the **Observation and Inferences** section of the Introduction to Physical Science Reading.

15   Astrolabe

Related National Science Education Standards:	
Teaching Standard A: inquiry-based	
Science Content:	systems, order, and organization
	evidence, models, and explanation
	constancy, change, and measurement
	evolution and equilibrium
	form and function
Science as Inquiry:	abilities to do scientific inquiry
	understanding about scientific inquiry
Physical Science:	properties of objects and materials
	position and motion of objects
	motions and forces
	light, . . .
Science and	
Technology:	abilities of technological design
	understanding about science and technology
	distinguish between natural objects and objects made by humans
Science in Personal and	
Social Perspectives:	changes in environments
History and Nature of	
Science:	science as a human endeavor
	nature of science
	nature of scientific knowledge
	historical perspectives

Exploration/Awareness:

In this course you are asked to carefully make observations of the Moon and to observe constellations. For most observations, estimating the **altitude** and **azimuth** will be sufficient. However, it is possible to make more careful measurements. An **astrolabe** is a device first used by the ancients to determine the altitude of celestial objects.

1. Use an already constructed astrolabe or construct your own astrolabe. A simple astrolabe can be constructed from a protractor, string or fishing line, washer, and drinking straw. Students can modify directions given by their instructor to construct a more sophisticated astrolabe.
2. From a location provided by your instructor, record the altitude and azimuth of three objects identified by your instructor in the table that follows.

Table 2-2   Altitude and Azimuth

Object	Altitude	Azimuth

Concept Development:

3. What would the altitude be of an object directly overhead?

4. What is the altitude of an object located about halfway up in the sky?

5. If north represents an azimuth of 0° (or 360°) and azimuth readings increase in a clockwise direction (when viewed from above), what is the azimuth of an object located in the south?

6. What is the azimuth of an object located in the southwest?

7. Can you determine the location of an object in the sky if you are only given the:
    a. altitude? Explain

    b. azimuth? Explain.

Application:

8. Use an astrolabe to record altitude and azimuth of the Moon while completing the Moon long-term observation activity.

Background Information:

For additional background, read the **Locating Objects in the Sky** section of the Introduction to Physical Science Reading.

Name(s)   _____

                 _____

                 _____

                 _____

16   Scientific Notation Practice

Convert the following to scientific notation and write your answer in the blank.

1. 100000   _____

2. 50.03   _____

3. 0.00983   _____

4. .247   _____

5. 3671800   _____

Convert the following from scientific notation to regular notation and write your answer in the blank.

6. 7.4×10^5 _____

7. 2.19×10^{-2} _____

8. 10^{-8} _____

9. 3.58×10^{-1} _____

10. 1.695×10^3 _____

For additional background, read the **Scientific Notation** section of the Introduction to Physical Science Reading.

17   SI Conversions Practice

Convert the following:

1.   716 km = _____ m

2.   9.0 L = _____ mL

3.   9721 g = _____ kg

4.   52.43 cm = _____ m

5.   10.0 kg = _____ mg

6.   1 gigabyte = _____ megabytes = _____ kilobytes

7.   65 mm = _____ cm = _____ m = _____ km

8. In general, describe the process when changing from a larger SI unit (not number) to a smaller SI unit (not number). Describe the process when changing from a smaller SI unit (not number) to a larger SI unit (not number).

For additional background, read the **SI Units of Measurement** section of the Introduction to Physical Science Reading.

18   Graphs and Graphing

Related National Science Education Standards:
Teaching Standard A: inquiry-based
Science Content: systems, order, and organization
evidence, models, and explanation
constancy, change, and measurement
evolution and equilibrium
form and function
Science as Inquiry: abilities to do scientific inquiry
understanding about scientific inquiry
Standards addressed will depend upon data graphed.

Prior to working on this activity you may need to read the **Graphs and Graphing** section of the Introduction to Physical Science Background Reading and/or receive instruction from your teacher.

Exploration/Awareness:
1. Use data provided or collect data as directed by your instructor to use to make and properly label two different types of graphs.

Concept Development:
2. In physical science courses, identify the requirements for a well-made graph.

3. Under what circumstances would it be proper to make a line graph? . . . a bar graph or histogram? . . . a pie chart?

Application:

4. Besides science and mathematics classes, identify two contexts where graphs are frequently used.

5. Find data that represent each of the patterns of change and graph the data.

6. After receiving instruction on the use of a graphing calculator, input data and graph the data.

7. Use other functions of the calculator to determine additional information about the graphs constructed with the use of the graphing calculator. Report the information in this space.

Background Information:

      For additional background, read the **Graphs and Graphing** section of the Introduction to Physical Science Reading.

19   Calibrating a Balance

Related National Science Education Standards:
  Teaching Standard A: inquiry-based
    Science Content: systems, order, and organization
                                     evidence, models, and explanation
                                     constancy, change, and measurement
                                     evolution and equilibrium
                                     form and function
    Science as Inquiry: abilities to do scientific inquiry
                                     understanding about scientific inquiry
    Physical Science: properties of objects and materials
                                       position and motion of objects
                                       motion and forces
    Science and
    Technology: understanding about science and technology
                                     Distinguish between natural objects and objects made by humans

Exploration/Awareness:

We don't often ask: "How much matter?" Frequently, we are concerned about the weight of various objects. Besides considering our own weight, what are some circumstances where people care about weight? Commonly, measurements of mass and weight are considered to be interchangeable. In science, a distinction is made between mass and weight.

A balance is an important scientific tool. You will use your equal-arm balance frequently so it is important for you to calibrate it with sufficient care. In the process of calibrating your balance, it is hoped that you will understand the way this particular measuring device works.

1. Make a sketch of your balance and label these parts: left and right arms, left and right pans, pivot point or balance point or fulcrum, left and right riders, and pointer.
2. With both pans attached, slide the right rider as far to the left as possible with the balance still able to move freely. Next, position the left rider until the pointer swings equal distances on either side of the midpoint of the scale. If the pointer is swinging wildly, you can slow it down by holding it near the midpoint, releasing it, and then making any necessary adjustments. If the balance does not adjust with the movement of the left rider, try reversing the pans. You may have to move the right side rider, but you should not move it too far to the right. Once the balance has been properly adjusted, squeeze the left side rider tightly to prevent it from sliding around too much and use the china marker pencil to make a mark where both left and right riders need to be located. These marks represent your balance point or "zero" point.

3. With the riders in the "zero" positions, your balance should swing equal distances to either side of the midpoint of the scale. Place a 100 milligram (mg) mass on the left pan. Move the right side rider along the arm until it balances, i. e., it swings equal distances to either side of the pointer scale. Mark the position of the right rider. Divide the distance between the right side "zero" mark and the right side "100 mg" mark into ten equal spaces and use the china marker pencil to mark each space. Each mark represents 10 milligrams (0.010 grams).

4. To make mass determinations, the object whose mass is **unknown** is placed in the **left pan** and **known** or standard masses are placed in the **right pan**. The right side rider may or may not have to be moved to get the system to balance. The **mass** of the **unknown** object is found by **adding** the **masses** in the **right pan** and **the mass represented by the position of the right side rider**. Care should be taken to insure that you are **adding** the **same units**, e. g., all the units are grams or all the units are milligrams and then find the sum.

Concept Development:

5. Students often say that the balance system works because "the weight is the same" on both sides. If the system is in balance, you can make it go out of balance by changing the position of one of the riders. Based on this, what two factors are involved in making determinations of mass?

6. What mass would you have needed to use when calibrating the balance so that each of the ten equal divisions would represent 100 mg?

7. What do the prefixes "kilo" and "milli" represent?

8. How many grams are in a kilogram? How many liters are in a milliliter?

Application:

9. Your balance swings equally with an unknown mass in the left pan and the following in the right pan: one 500 g mass, two 200 g masses, one 100 g mass, two 1 g masses, one 200 mg mass, and one 50 g mass. In addition, the right side rider is located at the 36 mg position. What is the mass of the unknown object?

10. How many ounces are in a pound? What is the weight in pounds of an object that weighs 9.6 ounces?

11. What is the mass in kilograms of an object that has a mass of 4768 grams?

12. In general, compare the process of converting in SI units and in converting in English or USCS units.

13. Explain how a teacher with limited resources could build a balance system using a meter stick, standard masses, string, and rubber bands.

14. What is the difference between weight and mass?

15. Does location affect the amount of mass? Explain.

16. Does location affect weight? Explain.

Background Information:

For additional background, read the **Mass and Weight** section of the Introduction to Physical Science Reading.

20   **Accuracy, Reproducibility, Sensitivity, and Uncertainty**

Related National Science Education Standards:
Teaching Standard A: inquiry-based
Science Content: systems, order, and organization
evidence, models, and explanation
constancy, change, and measurement
evolution and equilibrium
form and function
Science as Inquiry: abilities to do scientific inquiry
understanding about scientific inquiry
Physical Science: properties of objects and materials
position and motion of objects
motions and forces
Science and
Technology: understanding about science and technology
Distinguishing between natural objects and objects made by humans

Exploration/Awareness:

When you buy something that is sold by a measured amount, do you ever wonder if you are actually getting the correct amount? Is a gallon of gas really a gallon of gas? Is a pound of hamburger really a pound of hamburger? Is a 16 ounce drink really 16 ounces?

1. Alternately and **without telling others your results**, have at least three different people in your group determine the mass of an aspirin tablet. Properly record the masses in the "Determining Mass" table found at the end of this activity.

a. Did each person get the same mass for the aspirin tablet? Explain your results.

b. Ask your instructor for the accepted value of the aspirin. Do you consider your measurement of the aspirin's mass to be accurate? Why or why not?

2a. Select a "light" object. **Without telling others your results**, have one member of your group find the mass of the "light" object. Properly record the mass in table found at the end of this activity.

2b. Again, without telling others your results, have a second member of your group find the mass of the light object and properly record the mass in the table.

2c. Have both persons find the mass a second time without telling anyone their results and properly record the measurements. After each has found the mass a second time, share the results so that other group members can properly record the masses in their data table.

2d. When an individual found the mass twice, did they get the same result each time? Explain why or why not.

2e. Did both individuals get the same mass for the light object? Explain why or why not.

3. Select a "heavy" object. **Without telling others your results**, have one member of your group find the mass of the "heavy" object. Properly record the mass in the table. Similar to what you did with the light object, have a second group member find the mass and properly record the mass. Have each get a second mass measurement, properly record their results, and share their results with other group members.

3a. When an individual found the mass of the heavy object twice, did they get the same result each time? Explain why or why not.

3b. Did both individuals get the same mass for the heavy object? Explain why or why not.

4. To determine the sensitivity or the smallest amount that will cause your balance to "go out of balance:"

4a. Cut out 100 squares of small grid graph paper and properly record the mass:

_____

4b. While in balance, cut off one square at a time and remove it/them from the balance pan until the balance system goes out of balance. What is the least number of square(s) that cause(s) the balance to "go out of balance?"

_____

4c. Use the relationship that follows to calculate the sensitivity by determining the mass of the squares that caused the balance to "unbalance." In this example, we will use x to represent the sensitivity.

$$\frac{\text{mass of 100 squares}}{\text{100 squares}} = \frac{x}{\text{number of squares to "unbalance"}}$$

x or the sensitivity is _____

Concept Development:

5. **Uncertainty** can be calculated in a series of steps. Start by calculating the average value for a measurement. Calculating the average value for a measurement is accomplished by simply adding the measurements together and dividing by the number of measurements. The next step in calculating uncertainty is done by determining the difference between the average value and each one of the measurements. Since we are concerned with only the numerical difference, not the direction of the difference, there will be no negative numbers. The "average of the differences," the uncertainty, is calculated by adding the differences and then dividing by the number of differences (measurements). An example is worked below.

a. Measurements of mass of an aspirin tablet (3 measurements):
   345 mg   346 mg   341 mg

b. Average value: 345 mg $\frac{1032 \text{ mg}}{3} = 344$ mg
                    341 mg
                    <u>346</u> mg
                 1032 mg

c. Differences between measured values and average value:

   345 mg – 344 mg = 1mg
   346 mg – 344 mg = 2 mg
   344 mg – 341 mg = 3 mg

d. "Average of the differences:"
        1 mg $\frac{6 \text{ mg}}{3} = 2$ mg
        2 mg
        <u>3 mg</u>
        6 mg

e. The measurement recorded with uncertainty would appear as: 344 mg $\pm$ 2 mg

f. The measurement can be recorded with uncertainty as a percent of the average value as illustrated in the following example:

$$\frac{344 \text{ mg} \pm 2 \text{ mg}}{344 \text{ mg}}$$

344 mg $\pm$ .0058139 which, written as %

344 mg $\pm$ .6%

6. In general, identify some potential sources of error when measuring.

Application:

7. What can be done to improve accuracy and reduce uncertainty?

8. Calculate the uncertainty in your aspirin tablet measurements. Show your work in the space below. Properly record the measurement with uncertainty and with uncertainty as a percent.

9. Write a statement that summarizes your knowledge of uncertainty by incorporating these three terms: accuracy, reproducibility, and sensitivity.

Background Information:

For additional background, read the **Measurements** section of the Introduction to Physical Science Reading.

Table 2-3   Determining Mass

Object	Measured Mass Trial 1	Measured Mass Trial 2	Measured Mass Trial 3	Average Mass
1. aspirin				
2a & c. "light" object by _____			X	X
2b & c. "light" object by _____			X	X
3. "heavy" object by _____			X	X
3. "heavy" object by _____			X	X

21   Meter Stick Balance

Related National Science Education Standards:
Teaching Standard A: inquiry-based
Science Content: systems, order, and organization
evidence, models, and explanation
constancy, change, and measurement
evolution and equilibrium
form and function
Science as Inquiry: abilities to do scientific inquiry
understanding about scientific inquiry
Physical Science: properties of objects and materials
position and motion of objects
motions and forces
Science and
Technology: abilities of technological design
understanding about science and technology
distinguishing between natural objects and objects made by humans

Exploration/Awareness:

In an earlier activity, you discovered that it is more than mass (or "weight") that makes a balance work. Most people have an intuitive feel for how a teeter-totter or see-saw works. A heavier person won't balance a lighter person if each one sits at the end. To balance people with unequal weights, the people must sit at different distances from the fulcrum. This activity will give you a chance to move beyond a descriptive level of understanding of how a balance works to the more abstract level where you will discover a mathematical relationship and write an equation that describes this relationship.

1. Utilize whatever balance system is available. For most this will probably be a meter stick with a hole drilled into it and an attached string from which it can be suspended, two rubber bands, some string, and a pegboard or ring stand to serve as a support for the meter stick. Make sure the meter stick is initially in balance, i. e., when suspended on your finger with a piece of string attached near the middle of the meter stick, the meter stick will not rotate in any direction.
2. Using combinations of 200 g, 100 g, and 50 g masses, suspend one mass on the left side of the meter stick and a different mass on the right side of the meter stick. Move the masses around until the system is in balance. SAFETY PRECAUTION: Don't let the meter stick hit you as it may move around as you make adjustments! In the Balance with Two Masses table at the end of this activity, record the mass on the left side as m_1 and record the mass on the right side as m_2. The **distance from the pivot**

point or fulcrum to where the mass is suspended is the length. Note the length or distance is not simply the centimeter reading where the mass is suspended.
Record the lengths, (l_1 as the distance from fulcrum to where m_1 is suspended and l_2 as the distance from the fulcrum to where m_2 is suspended) in the data table.

3. Do enough trials until you can successfully predict, within certain ranges, combinations of masses and distances from the fulcrum that will balance the system.

4. Repeat the balancing procedure with three masses instead of two. Number the masses and distances left to right as 1, 2, and 3. Record the data in the Balance with More than Two Masses table.

5. Do enough trials until you can successfully predict combinations of masses and distances from the fulcrum that will balance the system.

Concept Development:

6. Based on data collected with two masses and two lengths, **use symbols** to develop a **proportion** and then an **equation** that illustrates the relationship between masses and lengths required to balance a system with two masses and two lengths. Your instructor can provide assistance if you need help. Write the equation out in **words**.

7. Based on data collected with three masses and three lengths, develop an equation that illustrates the relationship between masses and lengths required to balance the system.

8. Write a statement that describes the conditions necessary for a system to be in balance.

9. Write a statement that describes the conditions necessary for a system to rotate in a counterclockwise direction.

10. Write a statement that describes the conditions necessary for a system to rotate in a clockwise direction.

Application:
11. A meter stick balance system has a fulcrum at the 50 cm mark. A 100 g mass is suspended at the 90 cm mark and a 200 g mass is suspended at the 30 cm mark. Make a sketch of the system. Is the system in balance? Explain why or why not.

12. A meter stick balance system has a 200 g mass suspended at the 75 cm mark and has a 100 g mass suspended at the 15 cm mark. The fulcrum is at the 50 cm mark. Make a sketch of the system. Where would a 50 g mass have to be placed for the system to balance?

Background Information:
    For additional background, read the **Torque** section of the Introduction to Physical Science Reading.

Table 2-4   Balance with Two Masses

m_1	m_2	l_1	l_2

Table 2-5   Balance with More than Two Masses

m_1	m_2	m_3	l_1	l_2	l_3

22   Make a Mobile

Mobiles are sculptures that can move and systems that involve torque. Construct a mobile with a physical science theme.

23   Introduction to Physical Science Questions and Problems

(You must show your work on all problems to receive credit!)

1. What factors determine whether a "balance system" is in balance?

2. Using a meter stick with a fulcrum at the midpoint, a student suspended a 50 g mass at the 15 cm mark and a 100 g mass at the 60 cm mark. Make a sketch of this system. Where would the student need to suspend another 50 g mass so that the system would be in balance?

3. Sketch a portion of a centimeter ruler and indicate a measurement that would illustrate the 10.05 cm as a properly recorded measurement.

4. Sketch a portion of a graduated cylinder and indicate a measurement that would illustrate 8.0 mL as a properly recorded measurement.

5. Sketch a portion of a graduated cylinder and indicate a measurement that would illustrate 8.00 mL as a properly recorded measurement.

6. The scale that follows represents lengths in centimeters. Read and properly record the lengths indicated by arrows a, b, and c.

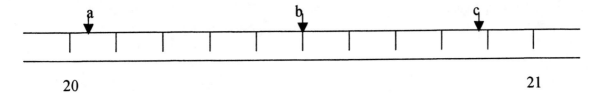

20   21

    a = _____

    b = _____

    c = _____

7. Ask nine of your classmates how many hours of sleep they average per night. Include information about yourself to make a tenth person. Record the data in a table. Construct and properly label a graph of the data.

8. Convert the following SI units:

a.   5.8 kg = _____ g

b.   49 mm = _____ m

c.   289 mL = _____ L

d.   0.006 km = _____ mm

e.   3 km = _____ cm

9. Convert the following from regular notation to scientific notation or vice versa.

a.   1000000 = _____

b. 9.9×10^{-2} = _____

c. 5.4×10^{4} = _____

d.   0.00001 = _____

e.   2345 = _____

f. $7.89 \times 10^3 =$ _____

10. Construct a concept map that incorporates the following science concepts: physical science, measurement, observation, experiment, and scientist.

11. Design a control variable experiment that investigates the effect of type of fabric (e. g., wool, cotton, polyester, etc.) and its insulating value.

12. Translate the following azimuth and altitude numbers into words that describe the general location of an object located at the azimuth and altitude numbers given.

azimuth = 180°   _____

altitude = 60°   _____

13. Which represents better vision 30/20 or 40/20? Explain.

Introduction

Various reasons are given in answering the question, "Why is science taught?" Project Synthesis (Harms and Yager, 1981) identified four practical reasons that can be remembered as the "CAPS" goals. Science is taught so that persons hoping to pursue **careers related to science** will have the needed science background. It must be noted that the percentage of high school seniors that eventually obtain careers in science is relatively small. Science is taught for **academic** reasons. This is the argument that science taught now will help you in science next year or on a future test or in college. In addition, Project Synthesis noted that science is taught for **personal** and **societal** reasons. Individuals can benefit from knowing science and society requires a scientifically literate citizenry to make informed decisions regarding scientific matters. In addition to reasons cited by Project Synthesis, other rationale can be cited for teaching science.

It is the nature of humans to seek knowledge. Some ways of knowing include: 1) **experience**; 2) **authority**; 3) **logical reasoning**; and, 4) **science**. We can know something by experiencing it. We can know something by having an authority figure, like a teacher or textbook, deliver the information. Deductive reasoning, reasoning from the general to the specific, and inductive reasoning, reasoning from the specific to the general, are other ways of knowing. Finally, science as a way of knowing will be explored in this textbook. Scientists, through various procedures and processes not exclusively their own, have developed knowledge. What are some of the advantages and disadvantages of the different ways of knowing? Can we experience all things? Can we trust all authorities?

The use of **science** and **technology** are human inventions that help societies **solve problems**. Societies and individuals can deal with problems in other ways. **Religion** may be helpful in dealing with ethical or moral problems. **Governments**, by making laws, or **society**, through the use of peer pressure for example, are modes for solving problems. Different problems may require different methods of solving problems. Many of today's problems are so complex that they require more than one approach. Pollution of the environment is a classic example of where science and technology, governments, ethics, and economics are interrelated.

There should be no doubt that we live in an increasingly complex world where individuals and societies can benefit from scientific knowledge. Scientific literacy is an often stated goal of many schools. Components of scientific literacy include: knowledge of science processes and products or facts, as well as understanding the relationship of science to society and technology.

Observations and Inferences

Observation involves use of the **senses** or an extension of the senses (e. g., using a microscope or telescope). Human senses include: **seeing, touching, tasting, smelling, and hearing**. Generally, humans rely on the sense of sight more than the other senses. **Resolution** is the degree to which details are distinguishable in an image. **Acuity** refers to the sharpness of an image. Astigmatism is faulty vision caused by defects in the lens of the eye that prevent focusing and getting a sharp, distinct image. Can you name an animal where sight is not the predominant sense?

An **inference** is a conclusion or an interpretation based on observations. Scientists frequently attempt to reconstruct events that occurred in the past or events that were not observed directly. Models or theories may be developed based on careful observations and inferences.

Science and Mathematics

Science and mathematics have long held a synergistic relationship. The use of mathematics has been integral to the tremendous advances made by science. Scientists use mathematics to analyze and describe concepts in nature. When science concepts are expressed mathematically, they are unambiguous. All scientific disciplines are bolstered when data are analyzed using mathematics. While recognizing the importance of mathematics to science, students should first understand science concepts before they prematurely attempt mathematical applications.

Scientific Notation

The speed of light in a vacuum is about 1.86×10^5 miles per second or 2.9979×10^8 meters per second.

A mole represents about 6.02×10^{23} particles.

A millimeter is 10^{-6} kilometers.

The average distance between the Sun and Earth is 1.5×10^8 km.

Listed above are examples of numbers written using scientific notation. Numbers that are extremely large or extremely small are awkward to write because it is difficult to keep track of all the digits. **Scientific notation** is an example of exponential notation or powers of ten notation. Scientific notation is a shorthand or condensed way of dealing with very large or very small numbers because it does not involve writing down a lot of zeros. A superscript is a number written to the right and above a letter or number. In algebra class you may have called a superscript an exponent. The superscript or exponent indicates how many times the letter or number is multiplied by itself. For example, x^2 means "x times x;" it does not mean "x times 2." Similarly, 10^4 does not mean 10 with four additional zeros, it means "10 x 10 x 10 x 10" or 10,000. Numbers are written in scientific notation when they are written as **10 with an exponent or 10 to some power** or when they are written as a **number between 1 and 10 times 10 to some exponent**. You may have used a calculator when doing a problem where the answer was reported in a form of scientific notation. Here are two possible ways that calculators may report an answer in a form of scientific notation.

1.89 10

1.89E10

In both cases the answer is another way of writing: 18,900,000,000. In the first example, blank spaces indicate that the answer is 1.89 times ten to the tenth power (10^{10}). In the second example, the "E" (for exponent) indicates that the answer is 1.89 times ten to the tenth power.

Listed below are some numbers written in powers of ten or scientific notation, as ten multiplied by ten whatever number of times the exponent indicates, and in "regular" notation.

$$10^3 = 10 \times 10 \times 10 = 1000$$
$$10^2 = 10 \times 10 = 100$$
$$10^1 = 10$$
$$10^0 = 1$$
$$10^{-1} = 1/10 = .1$$
$$10^{-2} = 1/10 \times 1/10 = 1/100 = .01$$
$$10^{-3} = 1/10 \times 1/10 \times 1/10 = 1/1000 = .001$$

Do you see the pattern? How would 1000000000 (one billion) be written using scientific notation? How would .000001 (one-millionth) be written using scientific notation?

How are numbers that are not powers of ten or ten to some exponent written in scientific notation? Recall, such numbers to be written in scientific notation must be written as a number between 1 and 10 times 10 to some exponent. Two examples follow:

Example 1:
1.   Write 5623.56 in scientific notation.
Step 1 - Move the decimal point to the right or left, left in this case, so the number will be between 1 and 10.

$$5\ 6\ 2\ 3\ .\ 5\ 6$$

In this case the decimal point would be moved 3 places to the left.

Step 2 – The number of places the decimal point is moved will become the exponent when the number is written in scientific notation.

$$10^3$$

Step 3 – The exponent is positive if the decimal point was moved to the left to write the number as a number between 1 and 10 and negative if the decimal point was moved to the right. Frequently, "+" is not written when the exponent is positive, but the "-" is written to indicate a negative exponent.

$$5.62356 \times 10^3$$

Example 2:

2.   Write .098714 in scientific notation.

   Step 1 -

.0 9 8 7 1 4

   Steps 2 and 3 –

9.8714 x 10^{-2}

Measurement

A **measurement** includes a **number** and a **unit** of measure. When recording a measurement the convention in this course will require you to always record a number and a unit of measure.

Most of us have made measurements. Think about the process that you go through when you make a measurement. First, you **decide** on the feature, characteristic, or **attribute** to be measured such as distance or length. Probably your second step is to select a unit that has that attribute. You can use a **standard unit** or a **non-standard unit**. A meter is an example of a standard unit to measure distance or length. Your foot may or may not be one foot long, i. e., twelve inches. What are the advantages and disadvantages of standard units versus non-standard units? Next, you may **compare** the unit to the attribute to be measured by filling, covering, placing side by side, . . . etc. or some other method. In the case of length or distance you might place the measuring unit along the distance or length to be measured. The distance or length may be a fraction of the unit or a multiple of the unit. Finally, there may be a need to **record** the measurement.

Scientists follow a particular way of reading and recording measurements. They **record** the **digits** that they can **read plus one more digit**. These digits, i. e., the digits that can be read plus one more, are considered to be **significant**.

One develops understanding of the attribute being measured by making comparisons based on the attribute. Making **estimates** prior to measuring is a useful step that gives the unit meaning. Using physical models of measuring units to fill, cover, or place side to side develops **understanding** of how this comparison process produces a measurement. **Making measuring instruments** and **using** them develops **understanding** how **measuring instruments work**.

Accuracy is how close a measurement is to the actual, accepted, standard, or known value or true value. **Precision** has to do with the range of scatter or uncertainty. Accuracy and precision are sometimes used interchangeably. Precision is associated with how easily results can be duplicated. A person who can hit a dart board bull's eye is accurate. If the person cannot repeatedly hit the bull's eye, the person is not precise. **Reproducibility** has to do with getting the same results each time. **Sensitivity** is a measure of the smallest amount that makes a difference. In the case of a balance, it would be the smallest amount that causes the measuring device to go out of balance. **Errors** associated with measuring can be attributed to either the **measuring instrument**

or to the **person** doing the measuring. Because there can be error in any measurement, some scientists like to think of a measurement as having some **uncertainty**. **Scatter** is a term associated with uncertainty and refers to the range of values obtained in the measurement process. In reporting a measurement, some scientists like to report the highest and lowest measured values along with the average value of the measurements.

SI Units of Measurement

Systems of units have been developed to measure how much of a given quantity there is. In addition, systems of units are an example of where governments are involved with science and commerce. When you buy a gallon of gasoline or a pound of meat, you want to make sure that you are getting the full amount for your money. Two major systems of measurement are used in the United States: the **United States Customary System** (**USCS** or **English** system) and the **Systeme International** (**SI**, and also known as the international system or the **metric system**). SI or metric units are used throughout most of the world. The SI system has the advantage of being a decimal system where units are related to smaller or larger units by dividing or multiplying by 10. In the United States, the SI system is rapidly replacing the USCS system where the SI system is more convenient to use, e. g., when products are intended for overseas sales and in some sports like track and swimming. The substitution is occurring more slowly in other areas and, in some cases, the changeover may never occur. Expenses, geography, and convenience are factors to be taken into consideration when the decision to switch from one measurement system to another is being made.

As stated earlier, a major advantage of the SI system is that it is a decimal system. A list of some SI prefixes are provided in the SI Prefixes Table that follows. For a long time the prefixes most commonly used were kilo, centi, and milli. However, prefixes such as giga, mega, micro, and nano are being used more frequently.

SI is based on the metric system which was developed by French scientists after the French revolution in 1791. The unit of **time** in both the SI and USCS systems is the **second** (s). Within the SI system, reference is sometimes made to the **mks system**. In this system, the unit of length is the **meter**, the unit of mass is the **kilogram**, and the unit of time is the **second**. Other times, reference may be made to the **cgs system**. The units are **centimeter**, **gram**, and **second** for length, mass, and time respectively. The mks and cgs systems differ in size; 1 meter equals 100 centimeters and 1 kilogram equals 1000 grams. The most familiar SI unit for most students in the United States is the **liter**, a unit for **volume**. Soft drinks are commonly sold in one- and two-liter bottles. A liter is the amount enclosed by a cube ten centimeters on a side (1000 cubic centimeters).

Time

Both SI and USCS systems use the **second** (s) as a unit of time. In SI, prefixes listed in the SI Prefixes Table can be used with second. In addition to second, in the USCS system, minute, hour, day, and year are used.

Length

The standard unit of length is the **meter**. At one time, the French Academy defined the meter as one ten-millionth of the distance between the equator and the North Pole at the longitude of Paris. During this time period, the distance was close to 10,000 kilometers. Today the meter is defined as the distance traveled by light in a vacuum during a time interval of 1/299,792,458ths of a second.

Figure 2-2   SI Prefixes[1]

Prefix	Symbol	Meaning	
giga	G	1,000,000,000	(10^9)
mega	M	1,000,000	(10^6)
kilo	k	1,000	(10^3)
hecto	h	100	(10^2)
deka	da	10	(10^1)
unit (either meter, m; liter, L; or gram, g)		1	(10^0)
deci	d	.1	(10^{-1})
centi	c	.01	(10^{-2})
milli	m	.001	(10^{-3})
micro	μ	.000001	(10^{-6})
nano	n	.000000001	(10^{-9})

[1]Note prefixes between giga and mega and micro and nano (because of their less frequent use) are not listed here but do exist.

Area

The unit of area is a square that has a standard unit of length as a side. In the cgs system it is **1 cm^2** and in the mks system it is **1 m^2**. The area of a given surface is determined by counting the number of square units that it would take to cover the surface. Formulas can be used to calculate the surface area of various regular geometric shapes. The area of a rectangle equals the base times the height. The area of a circle is equal to πr^2.

Volume

Volume refers to the amount of space an object occupies. The unit of volume is the space taken up by a cube that has the standard unit of length for its edge. In the cgs system it is $1\ cm^3$ and in the mks system it is $1\ m^3$. The volume of a given object is determined by counting the number of cubes that it would take to occupy the same amount of space the object occupies. Under certain conditions, **1 g of water = 1 mL of water = 1 cubic centimeter of water ($1\ cm^3$).**

Mass

Mass refers to the **amount of matter**. In the cgs system, the standard unit of mass is the **gram** and in the mks system the standard unit of mass is the **kilogram**. A gram is the mass of 1 cubic centimeter ($1\ cm^3$) of water at a temperature of 4° Celsius. The kilogram equals 1000 grams.

Mass and Weight

Technically, scientists consider mass and weight to be different. Loosely speaking and in many ordinary, everyday contexts, people use the terms interchangeably. If something is heavy, we say that it has a lot of matter. Scientists refer to **mass** as the **amount of matter** and **weight** as the **force due to gravity**. Weight can vary depending upon location. Mass does not depend on location. Mass is not the same thing as volume. SI units of mass include grams and kilograms. A USCS unit of mass is the not well known **slug**. The SI unit of weight is the **newton**. USCS units of weight include **pounds**.

Graphs and Graphing

Graphs illustrate **relationships** between two **variables** or among more than two variables. Typically, at least one of the variables or factors is quantitative, i. e., involves numbers. A properly constructed and labeled graph should allow the viewer to make sense of the graph without additional information. A graph should have a title and each axis should be labeled with the variable and the units for each variable.

There are different types of graphs: **line graphs**, **bar graphs**, and **pie graphs** or **charts** are examples. It makes sense to use line graphs when the data are continuous and bar graphs or pie graphs for discontinuous data. Sometimes data do not fall exactly on a straight line. In some cases it might be appropriate to draw a **best fit line**. Although there are mathematical procedures that can be used to draw a best fit line, in this course it will be sufficient to sketch this line by observation without resorting to mathematical procedures. Bar graphs are frequently used when one variable represents discrete data. Eye color is an example of discrete data. Eye color is viewed as being brown or blue, for example. Pie graphs or charts are useful when representing fractional parts of a whole. Perhaps in a group of students 40% have blue eyes, 50% have brown eyes, and 10% have green eyes. To represent this distribution, a circle can be divided into parts that represent 40%, 50%, and 10%, respectively, of the circle. The choice of which type of graph to use depends on the type of data and consideration of which type of graph most effectively communicates information to readers.

The usual procedure in science is to place the **dependent** or **responding variable** on the ordinate or y-axis. This axis is the one that runs up and down or vertically. The

independent or **manipulated variable** is placed along the abscissa or x-axis. This axis runs side to side or horizontally.

The American Association for the Advancement of Science (1993) identifies three patterns of change that may be illustrated using graphs: **steady trends**, **cycles**, and **random**, **irregular**, or **no pattern**. Steady trends are illustrated when, as one variable increases, the other variable increases or decreases proportionately. Cycles illustrate a pattern that repeats. Random, irregular, or no pattern are terms that are descriptive of the lack of a pattern. Either no pattern exists or the pattern is too complex to be recognized. A **direct relationship** is one where as one variable increases (or decreases) the other variable increases (or decreases) proportionately. An **inverse relationship** is illustrated as one variable increases (decreases) the other variable decreases (increases) proportionately.

Occasionally, data can be collected continuously. More frequently, data are collected over distinct time intervals. In such cases time exists before the first data are collected and time after the last data are collected. In addition, time exists in between points where data are being collected. If a pattern of change exists, **interpolation** allows one to determine or estimate values in between known values. **Extrapolation** allows one to estimate or determine values before the first data collected or after the last data collected.

Locating Objects in the Sky

There are two systems for locating objects in the sky that will be used in this course. The first, the **altitude** and **azimuth** system, has the advantage of being easy to use. This system is useful in the vicinity of the observer's location but it is a disadvantage that altitude and azimuth measurements would be different for observers in other geographic locations. The second system, **right ascension** and **declination**, will be discussed and used during the unit on earth science.

Altitude in this context refers to **height above the horizon** measured in **degrees**. The point directly overhead, i. e., at 90°, is called the **zenith**. Altitude can range from 0° (at the horizon) to 90°. Azimuth refers to horizontal angular distance from a fixed reference point. North will described as being either 0° or 360°, east at 90°, south at 180°, and west at 270°.

In earlier times, humans more closely observed the motions of the Sun, Moon, planets and stars. Note the following origins, with slight variations depending on the source, for the words we use to name the days of the week:

Sunday – from Middle English and Old English, *sunnandaeg*, "day of the sun"

Monday – from Middle English and Old English, *monandaeg*, "moon's day"

Tuesday - from Middle English and Old English, "day of Tiu," from Germanic mythology Tiu was the god of war and sky, identified with the Norse god Tyr

Wednesday - from Middle English and Old English as a translation of the Latin *mercurii dies*, "day of Mercury"

Thursday - from Middle English and Old English, influenced by Old Norse, translation of the Late Latin *Jovis dies*, "Jupiter's day"

Friday - from Middle English and Old English, translation of Latin *Veneris dies*, "Venus' day"

Saturday - from Old English, *saeternesdaeg*, "Saturn's day"

Models

Some physical science phenomena can be observed directly but others cannot. When they cannot be observed directly or as a convenience to simplify complex phenomena, **models** may be used. Models are used in physical science to illustrate one or more important features of physical science phenomena. Models may be physical or mental representations. Scale models can be used to represent real objects at a smaller size or a larger size. A globe is a scale model of Earth. A physical scale model is constructed at a fixed ratio between the size of the model and the size of the real object. Models change as knowledge about phenomena change and as the technological means to represent models change. The use of improved models over time hints at the important role that creativity and logic play in science. Also, it suggests that science is a work in progress.

Torque

More than mass is involved when using a balance to determine mass or in using a simple balance made by using a meter stick. **Mass** and **the distance the mass is from the fulcrum** are both factors that have important roles in balancing the system. If the system is not balanced, it will turn in a clockwise direction because the clockwise mass times length exerts a greater force than the counterclockwise mass times length and causes the system to rotate in a clockwise direction. Similarly, if the counterclockwise mass times length is greater than the clockwise mass times length, the system will rotate in a counterclockwise direction. The combination of mass times length is called "**torque**" or "**turning effect**." When a system is balanced, counterclockwise torque equals clockwise torque.

Describing Science

Science is difficult to define. "Science" is derived from a Latin term, *scientia*, and other terms which translate as "to know" or "knowledge." Science involves the study of **natural phenomena**. It involves **observation** and **verification**. Frequently, verification involves conducting **experiments**. Many aspects of science are considered **empirical**, i. e., relying on or derived from observation or experiment. Science assumes that nature operates **logically**, i. e., in nature things happen for a reason, and that scientists seek the **truth**. Over time, the public nature of scientific processes moves science closer to the truth. This process may progress relatively quickly or take hundreds of years. Characteristics and methods of science are used to help describe the **nature of science**. The nature of science will be discussed in some detail in this course.

It was in western Europe during the 16th and 17th centuries that the modern scientific way of looking at the world appeared. This is not written to diminish contributions of earlier individuals or other cultures. The scientific work done today builds on scientific contributions of the past. Science is unique in its near unanimous worldwide acceptability. Disagreements in science occur but, typically, they are resolved fairly quickly in a historical sense. Normally, scientists from all over the world can agree on what constitutes important scientific research.

Duschl (1990) argues that a main **goal** of science is the **development of scientific theories** which lead to **scientific understanding**. Accepted scientific theories have lots of evidence to support them. A scientific theory is not simply imaginary speculation. When making explanations, scientists avoid **anthropomorphism** and **teleological** explanations. Anthropomorphism is attributing human characteristics, behavior, or motivation to non-living objects, animals, or natural phenomena. Teleological explanations attribute design or purpose in natural phenomena. An example of a teleological explanation would be that a hot air balloon rises because hot air wants to rise. A hot air balloon system rises when the overall density of the balloon system is less dense than the surrounding air. Also, scientists do not rely on divine intervention to explain phenomena. Duschl (1990) classifies theories into three groups: **center, frontier,** and **fringe**. Center level theories have the most solid observation base, offer a core explanation, and have no competing theories. By contrast, fringe theories have many competitors and represent a new explanation. Fringe level theories do not represent the mainstream of scientific thought. Over time as the observation base becomes more solid, a fringe level theory may move into the frontier category. A frontier level theory has a good data base with established explanations and a few competitors.

Science can be taught as an integrated course or broken down into the disciplines of biology, chemistry, earth-space science, and physics. Biology is classified as a life science (focus on living things) and chemistry, earth-space science, and physics are classified as physical sciences (focus on matter and energy). The physical sciences consist of chemistry, earth science, and physics. **Chemistry** studies the composition, structure, properties, and reactions of matter. Earth science includes several sciences that study the origin, structure, and physical phenomena of Earth. Physics studies matter and energy and interactions between matter and energy.

The American Association for the Advancement of Science (1993) describes the nature of science by discussing a scientific worldview, scientific inquiry, and the scientific enterprise (p. 3-21). A scientific view of the world recognizes the natural world as capable of being understood, yet science cannot provide complete answers to all questions. Questions exist that do not fall under the realm of study through science. While scientific knowledge is considered durable, it is subject to change. What must happen for scientific ideas to change? Science is considered to be: universal with broad applications, empirical, and reproducible. There is a desire for quantification in science. Scientists attempt to identify and avoid biases. Science is not considered to be authoritarian but scientists with national and international reputations do exert influence on science that lesser scientists do not have. Scientists use logic and imagination to gather evidence in order to make explanations and predictions. Science is a complex social activity that has generally accepted ethical principles involved in the conduct of science. Individuals, universities, private industries, and governments may carry on scientific research.

Science and Society

Earlier, it was noted that science plays a role in helping humans solve problems. Science is a product of the human desire to understand. Science and society have always impacted one another. Products of science have created benefits and problems for society.

Scientific advances bring possibilities for improving the quality of life in some respects and threatening society in other ways. Medical science is one of the most obvious examples of how science has contributed to the quality of life. Unanticipated problems also develop from the use of scientific discoveries. Use of nuclear energy, certain chemicals, and genetic engineering have been beneficial but problems associated with these and other applications of science have caused some to question the value of science.

Science has always impacted society. Science and society have become increasingly interdependent over time. Changes in both science and technology have increased the rate that scientific knowledge accumulates and increased public access to news of scientific discoveries and problems. Science depends on society for support to conduct its activities. Society relies on science to solve problems and improve the quality of life for humans.

The Scientific Method

Many scientists argue that there is no one scientific method, rather there are characteristics and methods of science. Regardless, many science textbooks will describe a scientific method. This procedure is one that is useful to determine cause and effect. Science **experiments** frequently have as a starting point **observations** made by one or more scientists. Scientists go through a process of **data analysis**, i. e., they will have thought deeply about observations that have been made. Sometimes texts combine the observation and data analysis steps of the "scientific method" as **stating the problem**. Formulation of a **hypothesis** is the next step followed by a **test** or **experiment** that is designed and completed to verify the validity or non-validity of the hypothesis. A hypothesis is sometimes called an "educated guess" or "tentative explanation" because deep thought is involved. A hypothesis is not a "wild guess." A **conclusion** may be determined based on the results of the test or experiment or additional study may be required before a firm conclusion can be made. Increasingly, there is a demand to apply results of a test or experiment to produce practical and/or economic benefits.

Scientists and Science Careers – General

In this course you will be given the opportunity to learn about specific scientists and science careers. Any listing of the most important historical figures includes a significant number of scientists. Some names are easily recognized by most people: Newton, Einstein, Galileo, Pasteur, and Darwin are examples. Perhaps some other names are less well recognized by most people, but still made important contributions: Bohr, Bacon, Lavoisier, and Fleming are examples. That a significant number of scientists are recognized for their important contributions is a testament to the value that society places on the important role science serves.

Science careers literally range from A, e. g., astronomer, to Z, e. g., zoologist. In addition to the many science occupations that exist, a number of occupations benefit from having some science background. A match between a science career and an individual's values, interests, and aptitudes is most likely to produce job satisfaction.

Science Case Study – Harnessing Power

Science and technology played a vital role in the Industrial Revolution. To understand the impact and nature of the Industrial Revolution, it is necessary to know how people lived and worked in the 18th century, to learn about their tools, and the distribution of material and energy resources. Prior to the 1800s, manufacturing typically was done on a small scale in homes using small machines powered by wind, running water, or muscle power. The invention of the steam engine and other machinery made it possible to use large factories powered by coal where chemical energy was converted to mechanical work. James Watt contributed to the improvement of steam engines. Steam engines were used in a variety of ways: to power ships and locomotives, move coal, and fuel machinery used in manufacturing. A factory system more efficiently produced goods than those produced in homes. The Industrial Revolution is considered to have started in Great Britain because science was applied for practical purposes, workers shifted from agriculture to factory jobs, and transportation by sea provided access to markets and resources.

Increased productivity is one of the key benefits of the Industrial Revolution but there also were drawbacks. The Industrial Revolution contributed to an increase in unhealthy working conditions and child labor. Conflicts between factory workers and owners contributed to differences in politics. The Industrial Revolution helped transform parts of the world from rural, agricultural societies to urban, industrial societies.

accuracy

altitude

area

average

azimuth

balance arm, rider, and pan

calibration/calibrating

concept map

conversion

demonstration

empirical

error

experiment

fulcrum

gram

gravity

Industrial Revolution

interpolation

liter

mass

mean (average)

meniscus

model

nature of science

patterns of change

physics

proportional thinking (if-then thinking)

random/irregular/no pattern

reproducibility/reliability

scatter

science understanding (4 levels)

scientific method

sensitivity

Systeme International (SI, metric)

teleological

torque (clockwise, counterclockwise)

uncertainty

units (standard, non-standard)

variable (dependent or responding, independent or manipulated, control)

volume

zenith

acuity

anthropomorphism

astronomy

axis (x, y, ordinate, abscissa)

balance

best fit line

chemistry

constellations

cycle

earth science

energy

estimate

extrapolation

graduated cylinder

graph

hypothesis

inference

linking words

long-term observation

matter

measurement

meter

Moon phases

observation

physical science

precision

pseudoscience

relationship (direct, inverse)

resolution

science (described or defined)

scientific law

scientific notation

steady trend

technology

theory

turning effect

USCS (United States Customary System)

weight

Name(s)   _____

_____

_____

_____

26   **Comparing Total Mass**

Related National Science Education Standards:	
Teaching Standard A: inquiry-based	
Science Content:	systems, order, and organization
	evidence, models, and explanation
	constancy, change, and measurement
	evolution and equilibrium
Science as Inquiry:	abilities to do scientific inquiry
	understanding about scientific inquiry
Physical Science:	properties of objects and materials
	properties and changes of properties in matter
Earth and Space Science:	properties of Earth materials
History and Nature of Science:	nature of science

Exploration/Awareness:

In a previous activity you have worked with the concept of "mass," i. e., the amount of matter, and have used a balance to determine the amount of matter. In this activity, you will look at what happens to mass in a closed system.

1. Fill a small plastic bottle (about 30-35 mL) about three-fourths full of water. If water is spilled onto to the outside of the bottle, use a paper towel to dry the outside.
2. Nearly fill the bottle cap with salt. You need to be able to move the cap around without spilling the salt.
3. Determine the mass of the bottle with water and the cap with salt all together. Do not find their masses separately. Properly record this measurement in Figure 3-1.
4. Without spilling, pour the salt into the water and cap the bottle. Shake the capped bottle so that the salt will dissolve. It is not necessary that all the salt dissolve but some of it should.
5. Predict what you think the mass is after having some or all of the salt dissolve. Properly record your mass prediction in Figure 3-1.
6. Determine the mass of the bottle, water, with some or all of the salt dissolved, and the cap. Properly record the mass in Figure 3-1.

Concept Development:

In comparing the mass before and the mass after the change, three possibilities exist:

a) the measurement of the masses are exactly the same.

b) the measurement of the masses are the same within experimental error, i. e., they differ by an amount that is smaller than the uncertainty.

c) the measurements of the masses are significantly different, i. e., the mass has increased or decreased by an amount that exceeds uncertainty.

Figure 3-1   Comparing Total Mass

Mass of entire system before dissolving (step #3):   _____

Predicted mass of entire system after dissolving (step #5):   _____

Mass of entire system after dissolving (step #6):   _____

7. Why do you think you found the mass of the bottle, cap, water, and salt all at once instead of individually and then finding the total mass by addition?

8. What do you conclude about the total mass before and after dissolving salt in water?

9. If you wanted to provide evidence that the salt "did not disappear" when dissolved, what could you do?

Application:

10. Design an experiment that determines what happens to the mass of ice when it melts making sure that the system is a closed system and variables are controlled.

Background Information:

For additional background, read the **Law of Conservation of Mass** section of the More Physical Science Reading.

27   **Comparing Masses and Mass of a Gas**

Related National Science Education Standards:	
Teaching Standard A: inquiry-based	
Science Content:	systems, order, and organization
	evidence, models, and explanation
	constancy, change, and measurement
	evolution and equilibrium
Science as Inquiry:	abilities to do scientific inquiry
	understanding about scientific inquiry
Physical Science:	properties of objects and materials
	properties and changes of properties in matter
Earth and Space Science:	properties of Earth materials
History and Nature of Science:	nature of science

Exploration/Awareness:

In this activity you will need to make careful measurements to determine the mass of a gas.

1. Collect the following materials: a snack-sized resealable plastic bag, plastic bottle (30-35 mL) not quite filled with water, one-half or slightly less than one-half of a seltzer tablet, and a balance.

2. Check to make sure the plastic bottle is clean and dry and make sure the resealable plastic bag has no holes. The water will eventually be poured from the plastic bottle into the plastic bag so it is very important that the plastic bag has no holes and can be sealed completely.

3. In one "massing," find the total mass of the: snack bag, bottle nearly filled with water (you need to be able to move the bottle around without spilling the water so do not fill the bottle), and the seltzer tablet (about one-half or a little bit less than one-half of a tablet). Properly record the mass in the Figure 3-2.

4. Without spilling any water, hold the snack bag and the plastic bottle with water upright in the snack bag.

5. Place the part of a seltzer tablet in the bag, but not in the water! Carefully seal the bag.

6. After making sure the bag is sealed and does not have any holes, tip the bottle so the water and tablet will mix with one another.

7. In Figure 3-2, record whether you think the mass of the "bag system" has increased, decreased, or stayed the same as that recorded in step #3.

8. Observe evidence of a change. Properly record the mass of the total system in Figure 3-2.

9. Unseal the bag and allow or force some of the gas to escape. Reseal the bag and find the mass of the system. Properly record the mass in the table that follows.

Figure 3-2   Comparing Masses and Mass of a Gas

Mass of snack-sized bag, plastic bottle with water, and part of seltzer tablet (step #3)

_____

Predicted change in mass from step #3 (step #7):   increase   decrease   stay the same

Mass of snack-sized bag, plastic bottle with water, and part of seltzer tablet
after mixing (step #8)

_____

Mass of snack-sized bag, plastic bottle with water, and part of seltzer tablet
after releasing some gas from the bag (step #9)

_____

Concept Development:
10. What evidence did you observe that would allow you to infer that a change took place?

11. How is the change that occurred in this activity different from the change that occurred when you combined salt with water?

12. How did the mass of the system before and after mixing compare? Was the result expected? Explain your results.

13. How did the mass after mixing compare with the mass after releasing some gas from the plastic bag?

14. What evidence is needed to suggest that gas has mass?

Application:

15. Suggest and briefly describe an activity that would illustrate the Law of Conservation of Mass.

16. Describe changes where new substances are formed that have different properties from the original substances.

Background Information:

For additional background, read the **Properties of Matter** and **Law of Conservation of Mass** sections of the More Physical Science Reading.

28   **Area and Volume**

Related National Science Education Standards:	
Teaching Standard A: inquiry-based	
Science Content:	systems, order, and organization
	evidence, models, and explanation
	constancy, change, and measurement
	form and function
Science as Inquiry:	abilities to do scientific inquiry
	understanding about scientific inquiry
Physical Science:	properties of objects and materials
	properties and changes of properties in matter
	structure and properties of matter
	interactions of energy and matter
Earth and Space Science:	properties of Earth materials
Science in Personal and	
Social Perspectives:	types of resources

Exploration/Awareness:

Observing matter in its various forms produces opportunities to think about attributes like color, amounts or size, orientation, symmetry, and shapes. When comparing amount or size, it is important to decide what attribute is being compared. In some earlier activities you focused your attention on mass. In this activity, your attention will be focused on the attributes of area and volume.

1. Illustrated below, an irregular shape has been placed over a piece of square grid paper. Determine the area of the irregular shape by counting the number of squares that the irregular shape covers.

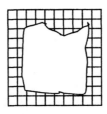

Area = _____ squares

2. Carpet is purchased in units of area. Identify the unit. List some other area units.

3. Gasoline is purchased in units of volume. Identify the unit. List some other volume units.

4. Some dry sand was poured into a 100 mL graduated cylinder to the 70 mL or 70 cm^3 mark. If you pour 30 mL of water into the graduated cylinder with sand, predict the total volume.

    predicted volume of sand water mixture = _____

5. A student found that when 30 mL of water was poured into the graduated cylinder with sand, the resulting volume of sand and water was 73.1 mL. Was your prediction accurate?

6. In another activity, a student poured 50 mL of water into one graduated cylinder and 50 mL of isopropyl (rubbing) alcohol into a different graduated cylinder. If combined, what do you think the total volume of the alcohol and water would be?

    predicted volume of alcohol and water mixture = _____

7. The student read the combined volume as 98.4 mL. Was your prediction accurate?

Concept Development:
8. Why isn't "length x width" a satisfactory definition for area in all situations?

9. List at least 3 different units for area.

10. A student remarked that multiplying length times width is like counting squares for area and multiplying length times width times height is like counting cubes for volume. Is this a reasonable conclusion? Describe why or why not.

11. How is volume determined for objects with irregular shapes?

12. List at least 3 different units for volume.

13. Provide a reasonable explanation for why the final volume of the sand and water mixture in #4 was not 100 cm^3 or 100 mL.

14. Provide a reasonable explanation for why the final volume of the water and alcohol mixture in #6 was not 100 cm^3 or 100 mL.

Application:
15. Sketch a portion of a graduated cylinder showing a measurement of volume that would properly be recorded as 8.66 mL.

16. From #4 and #5, determine the actual volume of the sand. Of the initial 70 mL or 70 cm^3, how much was space occupied by air?

17. For simplicity, assume an organism has a cube shape. If its length doubles, what happens to its volume (remember: the organism must retain its cube shape)? What does this suggest about what must happen to bones to support the increase in mass or weight associated with this increase in volume? Does this suggest a size limit on organisms?

18. A mineral with a volume of 22 cm^3 has a mass of 48.400 g.
    a. What is the mass of one cubic centimeter of this mineral?

    b. What would be the mass of another mineral (composed of the same material) having a volume of 500 cm^3?

19. With regard to size, a student asks: "Which is bigger?" What would be a better question?

Background Information:
    For additional background, read the **Area and Volume** section of the More Physical Science Reading.

29   Densities of Solids

Related National Science Education Standards:
Teaching Standard A: inquiry-based
Science Content: systems, order, and organization
evidence, models, and explanation
constancy, change, and measurement
form and function
Science as Inquiry: abilities to do scientific inquiry
understanding about scientific inquiry
Physical Science: properties of objects and materials
properties and changes of properties in matter
structure and properties of matter
interactions of energy and matter
Earth and Space Science: properties of Earth materials
Science in Personal and
Social Perspectives: types of resources

Exploration/Awareness:

Children, and some adults, will make references to one object being "heavier" than another object. "Heavy" and "light" are references to total mass and mass is an accidental property of an object not an intrinsic property. What is not being made clear in these comparisons is the need to compare "equal amounts" of the different substances. In this case, the equal amounts are equal volumes. A meaningful comparison is one that compares the masses of equal volumes.

In this activity, you will find the density of two regular shaped solids (one showing the density with percent uncertainty) and two irregularly shaped solids. Recall that $D = \dfrac{M}{V}$. Density units in this activity could be g/mL or g/ cm^3.

1. Use the Densities of Solids Table to conveniently organize your measurements and calculations. The table has the following headings: object, mass, volume, density, and density with percent uncertainty. Calculate the density of one of the regular shaped objects *with percent uncertainty*.

2. Select a regular shaped object and use a balance to measure its mass. Properly record the measurement in the table. If this is the object that you are going to use to calculate density with percent uncertainty you will need to: 1) take multiple measurements, 2) determine average mass, 3) calculate the differences between the average mass and the measured values, 4) calculate the "average of the differences,"

and then 5) calculate the uncertainty as a measurement and as a percent. Show your work! Space is provided after Table 3-1.

3. Determine the volume of the rectangular solid (regular shaped object) using L x W x H. Remember if this is the object that you are going to use to calculate density with percent uncertainty, you will need to: 1) take multiple measurements, 2) determine average volume, 3) calculate the differences between the average volume and the measured values, 4) calculate the "average of the differences," and then 5) calculate the uncertainty as a measurement and as a percent. Properly record the measurement in the data table. Show your work!

4. Use the formula to calculate density: density = average mass ÷ average volume. Properly record this in the data table. If this is the object you are using to determine percent uncertainty, add the percent uncertainty for the mass to the percent uncertainty for the volume.

5. Determine the mass and volume of a second regular shaped object and calculate its density. Properly record this information in the data table. Remember, if you earlier calculated percent uncertainty you do not need to do the calculation again unless you want more practice.

6. Determine the mass, volume, and density of two objects with irregular shapes and properly record the information in the data table. Select one sample of basalt and one sample of granite as the irregular objects. You will be using water displacement to obtain the volumes so make sure the pieces fit into a graduated cylinder.

Table 3-1   Densities of Solids

Object	Mass	Volume	Density	Density with % Uncertainty
regular object 1				
regular object 2				X
irreg. object 1				X
irreg. object 2				X

Space for calculations:

Concept Development:
7. You calculated density with percent uncertainty for one of the solids. Suggest a reason why the uncertainties for the mass measurement and the volume measurement are added.

8. Under certain conditions, the density of water is 1.00 g/cm^3. How do the calculated densities of the objects compare to water? Were any of the densities the same?

Application:
9. How could density be used to determine if two objects were composed of the same or different material?

10. How does the density of styrofoam "packing peanuts" explain why they work so well as packing material? Suggest a good reason why "packing peanuts" have the shapes they do instead of cubes or rectangular solids.

11. How could you find the density of an irregular object that floats?

Background Information:
    For additional background, read the **Density** section of the More Physical Science Reading.

30   **Densities of Liquids**

Related National Science Education Standards:
  Teaching Standard A: inquiry-based
    Science Content: systems, order, and organization
                                        evidence, models, and explanation
                                        constancy, change, and measurement
                                        form and function
    Science as Inquiry: abilities to do scientific inquiry
                                        understanding about scientific inquiry
    Physical Science: properties of objects and materials
                                        properties and changes of properties in matter
                                        structure and properties of matter
                                        interactions of energy and matter
    Earth and Space Science: properties of Earth materials
    Science in Personal and
      Social Perspectives: types of resources

Exploration/Awareness:

Recall that "heavy" and "light" are references to total mass and mass is an accidental property of an object, not an intrinsic property. A meaningful comparison is one that compares the masses of equal volumes.

In this activity you will find the density of two liquids. Recall the formula for density and units for density.

1. Use the Densities of Liquids Table that follows to conveniently organize your data. The table has the following headings: liquid, mass, volume, and density. Use water as one of the liquids and after calculating the density of water, select one other liquid from the choices provided. Space for calculations is provided after the table.
2. Use a graduated cylinder to carefully measure 10 mL of the liquid. Properly record this volume.
3. To obtain the mass of the liquid, you will first have to find the mass of a container, add the liquid to the container, find the mass of the liquid plus container, and then subtract to get the mass of the liquid only.
4. Calculate density. Properly record this in the data table. Show your work in the space provided.

Table 3-2   Densities of Liquids

Liquid	Mass	Volume	Density
water			
liquid other than water			

Space for calculations:

Concept Development:

5. Under certain conditions the density of water is 1.00 g/cm^3 or 1.00 g/mL. How does your calculated value for the density of water compare to the accepted value? List at least two reasons why your calculated value might differ from the accepted value.

6. How does the calculated density of the other liquid compare to water?

7. Write two general statements that describe the conditions necessary for objects to sink or for objects to float.

Application:

8. With safety in mind, use liquids of varying densities and food coloring to construct a "colorful visual display" that illustrates liquids of differing densities.

9. Place various cans of diet and regular pop (soda) in an aquarium containing water. Explain the results.

10. From prior activities you have evidence to support the Law of Conservation of Mass. References report the density of liquid water at 0° C is .9998 g/cm^3 and ice at the same temperature has a density of .9168 g/cm^3. If water freezes, the volume of the ice is how much greater than the volume of the water?

11. Design and conduct an experiment that answers the question: What is the effect of temperature on the density of a fluid?

Background Information:
For additional background, read the **Density** section of the More Physical Science Reading.

31   Density of a Gas

Related National Science Education Standards:	
Teaching Standard A: inquiry-based	
Science Content:	systems, order, and organization
	evidence, models, and explanation
	constancy, change, and measurement
	form and function
Science as Inquiry:	abilities to do scientific inquiry
	understanding about scientific inquiry
Physical Science:	properties of objects and materials
	properties and changes of properties in matter
	structure and properties of matter
	interactions of energy and matter
Earth and Space Science:	properties of Earth materials
Science in Personal and	
Social Perspectives:	types of resources

Exploration/Awareness:

In this activity you will find the density of a gas. In an earlier activity you determined that measuring the mass of a gas can be problematic.

1. Use Figure 3-3 Determining Mass of a Gas to conveniently organize your data. The table has the following headings: gas, mass, volume, and density.
2. You may have access to science equipment specifically designed to collect a gas. If not, use the following materials: plastic bucket or tub, 2 rubber bands, a paper clip, quart jar or similar container, large test tube, one-hole rubber stopper to fit large test tube with glass tubing inserted into the stopper opening, rubber tubing attached to the glass tubing, and a seltzer tablet broken in half.
3. Use a graduated cylinder to measure 10 mL of water and pour it into the test tube. Use the rubber band and/or paper clip to attach the test tube to a balance. Place the two halves of the seltzer tablet on the balance pan and find the mass of the "system" – the rubber band and/or paper clip, test tube, water, and broken seltzer tablet. Properly record this mass in Figure 3-3.
4. Put a rubber band around the quart jar.
5. Nearly fill the bucket or tub with water but allow enough space so you can place the quart jar in the water in the bucket or tub. Carefully submerge the jar so that you do not trap a lot of air bubbles in the jar.
6. Carefully turn the jar upside down (opening is down) so that no air bubbles are trapped in the jar and it remains filled with water.

7. Slide the end of the rubber tubing (not attached to the glass inserted in the stopper) so the end of the tubing is near the top of the collecting jar. This way the gas does not have to bubble through too much water. **You must not hold the jar down so tightly that the gas is unable to escape the test tube. This is a potential safety problem!**

8. If you have found the mass in step 3 and have made arrangements to collect the gas, you are ready to use the pieces of seltzer tablet and water to produce the gas. You will need to quickly drop the pieces of seltzer tablet into the water in the test tube and then, just as quickly, place the stopper in the test tube being careful not to pull the end of the rubber tubing out of the jar. The only place for the gas to escape should be through the rubber tubing into the jar to be trapped by the water in the bucket.

9. Allow the reaction to continue for 10 minutes. Place the test tube in a test tube stand so that you or your lab partners do not have to hold it.

10. After 10 minutes, use the rubber band to mark the volume of the gas collected in the glass jar. The rubber band should mark the gas/water line inside the jar. Carefully remove the rubber stopper from the test tube and the attached rubber tubing from inside the glass jar.

11. **Find the mass as before (but recognize this measurement does not include the mass of the gas! The gas was allowed to escape to the jar). Properly record this measurement in the Figure 3-3, Determining Mass of a Gas. Subtract the two masses in the in the table to obtain the mass of the gas! Do you understand why the difference in masses is recorded as the mass of the gas?**

Figure 3-3   Determining Mass of a Gas

Mass of the system before (#3):	_____
Mass of the system after (#11):	_____
Mass of gas (mass before – mass after):	_____

12. Record the mass of the gas in Figure 3-4.

13. Use a graduated cylinder to measure the amount of water needed to fill the jar to where the rubber band marks the volume of the gas. Properly record this measurement in Figure 3-4.

14. Calculate the density of the gas and properly record it. Show your work in the space provided after Figure 3-4.

15. Why was the seltzer tablet broken in half before you found the mass?

Figure 3-4   Density of a Gas Data Table

Gas	Mass	Volume	Density

Space for calculations:

Concept Development:

16 Assuming the densities of the materials you found are typical for solids (activity 29), liquids (activity 30), and a gas (this activity), make a general statement about how the densities of solids, liquids, and gases compare.

17. Use your own knowledge or a table of densities such as that found in the "Using Graphs" activity to identify some exceptions to your general statement developed in #16.

Application:

18. Identify 3 "real-world" examples of density that affect people.

19. Mass is added when helium is used to inflate a balloon, i. e., the amount of mass in the "balloon system" is increased. If released, the balloon would rise in the atmosphere. Explain how this happens.

20. Keeping in mind the formula for determining density and the Law of Conservation of Mass, explain how the average density of a hot air balloon changes.

21. Use the table of gas densities in the "Using Graphs" activity to identify the gas produced.

22. You can produce carbon dioxide gas by combining vinegar and baking soda. Put about 50 mL of vinegar in a plastic pop bottle. Put about 30 mL of baking soda into a balloon. Securely attach the balloon to the top of the pop bottle. Lift the end of the balloon up and shake it gently so that the baking soda mixes with the vinegar in the bottle. The balloon should inflate. After the balloon has inflated, take it off the pop bottle and quickly tie it off before the carbon dioxide escapes. What could you do to provide evidence that the gas produced is carbon dioxide?

Background Information:

For additional background, read the **Density** section of the More Physical Science Reading.

Name(s) _____

_____

_____

_____

32   A Real-World Example of Density: Wood

Related National Science Education Standards:

Teaching Standard A: inquiry-based

Science Content:	systems, order, and organization
	evidence, models, and explanation
	constancy, change, and measurement
	form and function
Science as Inquiry:	abilities to do scientific inquiry
	understanding about scientific inquiry
Physical Science:	properties of objects and materials
	. . . , heat, . . .
	properties and changes of properties in matter
	structure and properties of matter
	interactions of energy and matter
Life Science:	characteristics of organisms
	matter, energy, and organisms in living systems
Earth and Space Science:	properties of Earth materials
	energy in the Earth system
Science in Personal and	
Social Perspectives:	types of resources
	. . . resources, and environments
	natural resources
History of Science:	science as a human endeavor

Exploration/Awareness:

You have determined the densities of solids, liquids, and a gas. In this activity you will measure the mass and volume of some wooden blocks, calculate their densities, and determine if the density of the wood is related to other characteristics of wood.

1. Measure the mass and volume of the wood samples assigned to you and properly record the measurements in the table that follows.
2. Calculate the densities of the wood samples. Show your work. Record the densities in the table.

3. Based on the calculated densities of the wood samples, do the samples represent the same type of wood? Explain.

Table 3-3   Wood Density

Sample	Mass (g)	Volume (cm³)	Density (g/cm³)
1			
2			
3			

Concept Development:

Information in the following table was obtained from the U.S. Forest Products Laboratory (note: information from other sources may vary slightly).

Table 3-4   Wood Type and Relative Density

Type of Wood	Relative Density
Ash	.60
Aspen	.38
Beech	.64
Birch, Yellow	.62
Elm	.50
Fir, Balsam	.36
Fir, Douglas	.48
Hickory, Shagbark	.72
Maple, Red	.54
Maple, Sugar	.63
Oak, Red	.63
Oak, White	.68
Pine, Pitch	.52
Pine, Red	.46
Pine, Southern Yellow	.55
Pine, White	.35

4. Based on the information presented in the table and the calculated densities of the wood samples, make a reasonable determination of the types of wood represented by the samples. Identify the sample and the wood type in the Wood Sample Identification figure that follows.

5. The same piece of wood measured in the winter and then measured in the summer was found to have a slightly different density. What could account for this besides measurement error?

Figure 3-5   Wood Sample Identification

Sample: _____	Type of Wood: _____
Sample: _____	Type of Wood: _____
Sample: _____	Type of Wood: _____

Application:

A cord is a unit for measuring an amount of wood. A cord of wood represents a stack of wood with a volume of 128 ft^3. A Btu or British thermal unit is a unit for measuring energy. It represents the amount of energy needed to raise the temperature of one pound of water one degree Fahrenheit. Information in the Heating Value of Wood table that follows was obtained from the U. S. Forest Products Laboratory (note: information from other sources may vary slightly).

Table 3-5   Heating Value of Wood

Type of Wood	Heat Content (million Btu/cord)
Ash	20.0
Aspen	12.5
Beech	21.8
Birch, Yellow	21.3
Elm	17.2
Fir, Balsam	11.3
Fir, Douglas	18.0
Hickory, Shagbark	24.6
Maple, Red	18.6
Maple, Sugar	21.3
Oak, Red	21.3
Oak, White	22.7
Pine, Pitch	18.0
Pine, Red	12.8
Pine, Southern Yellow	14.2
Pine, White	13.3

6. Use data from the Wood Type and Relative Density table and the Heating Value of Wood table to construct a graph with relative density as the independent or manipulated variable and the heat content as the dependent or responding variable.

7. According to the graph, how is the relative density of wood related to its Btu content per cord?

8. Generally, what type or types of wood (or classification of wood) have high densities? What type or types have low densities?

9. If you were using wood to heat your home, what other qualities or characteristics of wood might be important considerations when purchasing wood for home heating?

10. Identify an example of density, other than the density example used in this activity, which affects people in daily life.

Background Information:
   For additional background, read the **Properties of Matter** and **Density** sections of the More Physical Science Reading.

33   **Sink, Float, Dive, Surface**

Related National Science Education Standards:
  Teaching Standard A: inquiry-based
    Science Content: systems, order, and organization
                                        evidence, models, and explanation
                                        constancy, change, and measurement
                                        evolution and equilibrium
                                        form and function
    Science as Inquiry: abilities to do scientific inquiry
                                        understanding about scientific inquiry
    Physical Science: properties of objects and materials
                                        position and motion of objects
                                        properties and changes of properties in matter
                                        motions and forces
                                        structure and properties of matter
    Earth and Space Science: properties of Earth materials
    Science and Technology: abilities of technological design

Exploration/Awareness:

In this activity, you will observe two different systems.

1. Either use a "diver system" that has already been constructed or construct your own diver system. Commercial materials may be purchased or you can use readily available materials. If you plan to construct your own diver system, you will need: 2L or 1L plastic pop bottle, beaker, and a medicine dropper.

2. Fill the beaker with water. Fill the dropper about half full of water and test it in the beaker of water. Adjust the amount of water in the dropper so that it just floats.

3. Fill the pop bottle with water, transfer the dropper from the beaker to the bottle, and with the bottle filled to the brim with water, screw the bottle cap on tightly.

4. The dropper or "diver" should float at the top. However, you should be able to make the dropper dive by gently squeezing on the sides of the bottle.

5. The second "system" will make use of a plastic cup, jar, or beaker and raisins, and a colorless pop or soda.

6. Fill the cup, jar, or beaker with the colorless pop. If this type of pop is not available, fill the cup with water and add a seltzer tablet.

7. Add a few raisins to the liquid. Over time, the raisins should bob up and down.

Concept Development:

8. What did the water level inside the dropper or diver do when it dove?

9. How did the volume of the air inside the dropper or diver compare when it was floating versus when it sank?

10. Is water or air more compressible?

11. What is the difference between the floating and sinking raisins?

12. Is there a difference between how long the raisins float at the top versus how long they sink to the bottom? Account for any differences.

Application:
13. Explain how the scientific principle demonstrated in this activity is related to submarines surfacing and submerging.

Background Information:
     For additional background, read the **Properties of Matter** and **Density** sections of the More Physical Science Reading.

34 **Viscosity Control Variable Experiment**

Related National Science Education Standards:	
Teaching Standard A: inquiry-based	
Science Content:	systems, order, and organization
	evidence, models, and explanation
	constancy, change, and measurement
	evolution and equilibrium
	form and function
Science as Inquiry:	abilities to do scientific inquiry
	understanding about scientific inquiry
Physical Science:	properties of objects and materials
	. . . and motion of objects
	light, heat, electricity, and magnetism
	properties and changes of properties in matter
	motions and forces
	structure and properties of matter
	interactions of energy and matter
Earth and Space Science:	properties of Earth materials
Science and Technology:	abilities of technological design
	understanding about science and technology
Science in Personal and	
Social Perspectives:	types of resources
	science and technology in local challenges
	. . . resources, and environments
	risks and benefits
	science and technology in society
	natural resources
History and Nature of	
Science:	science as a human endeavor

Exploration/Awareness:

You have made observations that allow you to state that not all fluids flow at the same rate. The ability to flow is a property of matter. **Viscosity** is the tendency of a fluid to resist flow. In comparing water and syrup, water has a lower viscosity because it flows easily and syrup has a higher viscosity because it resists flow more than water resists flow. In this activity, you are asked to design and conduct a control variable experiment that investigates some aspect of viscosity.

Materials needed will depend on the experiment you design. A suggestion is that you place the liquids on wax paper to assist with cleanup. Your instructor may provide the fluids. Once you have determined your procedure and the materials needed to complete the experiment, check with your instructor to see if the materials are readily available.

1. Write a **statement of the problem** or the **question** being researched:

Problem Statement or Question:

2. Write your **hypothesis**:

Hypothesis:

3. Briefly record your **procedure**:

Brief Procedure:

Concept Development:

4. **Communicate** the **results**: Use graph paper to illustrate the relationship, if any, between manipulated (independent) and responding (dependent) variables. Properly title and label the graph.

5. Write your **conclusion**:

Conclusion:

Application:

6. What factors may affect the viscosity of a fluid?

7. How could you use the viscosity rating of oil used in automobile engines to determine which would be best for you to use?

8. Identify three different properties of materials and illustrate how these properties affect the use of the materials.

Background Information:

      For additional background, read the **Properties of Matter** section of the More Physical Science Reading.

35   Strength and Elasticity Control Variable Experiments

Related National Science Education Standards:
Teaching Standard A: inquiry-based

Science Content:	systems, order, and organization
	evidence, models, and explanation
	constancy, change, and measurement
	evolution and equilibrium
	form and function
Science as Inquiry:	abilities to do scientific inquiry
	understanding about scientific inquiry
Physical Science:	properties of objects and materials
	. . . and motion of objects
	properties and changes of properties in matter
	motions and forces
	structure and properties of matter
Earth and Space Science:	properties of Earth materials
Science and Technology:	abilities of technological design
	understanding about science and technology
Science in Personal and Social Perspectives:	types of resources
	science and technology in local challenges
	. . . resources, and environments
	risks and benefits
	science and technology in society
	natural resources
History and Nature of Science:	science as a human endeavor

Exploration/Awareness:

You have made observations that allow you to state that not all materials have the same strength. **Strength** can be defined as a material's ability to resist changes in shape or breaking. In this activity, you are asked to design and conduct a control variable experiment that investigates the strength of uncooked spaghetti. **Elasticity** is the ability to return to its original state after undergoing deformation. To further demonstrate your ability to design and conduct a control variable experiment, you will design and complete a control variable experiment that tests elasticity.

Materials needed will depend on the experiment you design. For example, in the past some students used a plastic sandwich bag with a bent paper clip to suspend masses, like pennies or washers, from strands of uncooked spaghetti with different thicknesses. Once you have determined your procedure and the materials needed to complete the experiment, check with your instructor to see if the materials are readily available.

1. Write a **statement of the problem** or the **question** being researched:

Problem Statement or Question:

2. Write your **hypothesis**:

Hypothesis:

3. Briefly record your **procedure**:

Brief Procedure:

Concept Development:
4. **Communicate** the **results**: Use graph paper to illustrate the relationship, if any, between manipulated (independent) and responding (dependent) variables. Properly title and label the graph.

5. Write your **conclusion**:

Conclusion:

Application:
6. What factors may affect the strength of materials?

7. What properties of materials would be important considerations in the manufacture of:

   a. airplanes?

   b. tennis balls?

   c. golf clubs?

   d. disposable diapers?

   e. automobiles?

   f. jewelry?

8. Given materials provided by your instructor, design and complete a control variable experiment following standard procedures.

Background Information:
    For additional background, read the **Properties of Matter** section of the More Physical Science Reading.

36   **Using Graphs**

Related National Science Education Standards:
Teaching Standard A: inquiry-based
Science Content: systems, order, and organization
evidence, models, and explanation
constancy, change, and measurement
Science as Inquiry: abilities to do scientific inquiry
understanding about scientific inquiry
Physical Science: properties of objects and materials
properties and changes of properties in matter
structure and properties of matter
Earth and Space Science: properties of Earth materials

Exploration/Awareness:

You have completed activities that involved the densities of solids, liquids, and a gas. In this two-part activity you will investigate densities and circles using graphs.

Table 3-6   Densities at Room Temperature and Atmospheric Pressure (g/cm^3)

Solids		Liquids		Gases	
Material	Density	Material	Density	Material	Density
aluminum	2.7	benzene	.90	air, dry	
brass	8.6	ethyl alcohol	.81	(@ 0 °C)	1.29 x 10^{-3}
copper	8.9	gasoline	.66-.69	(@ 20 °C)	1.21 x 10^{-3}
glass	2.4-2.8	glycerin	1.26	ammonia	.77 x 10^{-3}
gold	19.3	mercury	13.6	carbon dioxide	1.84 x 10^{-3}
ice	.917	seawater	1.03	helium	
iron	7.85	water	1.00	(@ 0 °C)	.178 x 10^{-3}
lead	11.3			hydrogen	
platinum	21.5			(@ 0 °C)	.089 x 10^{-3}
quartz	2.6				
salt (table)	2.2				
silver	10.5				
tin	7.3				
wood – balsa	.11-.14				

1. You will use some density information to make a graph. Title your graph: Density of Selected Materials. Label the x-axis as volume (cm^3) and the y-axis as mass (g).
2. Use water and select three other substances that will make the graph relatively easy to make.
3. Plot the number of grams of each of the four substances in one cm^3. This gives you one point on what will be a line graph for each of the substances. Use proportional thinking to determine the number of grams in a volume other than one cm^3. For example, if you know the number of grams in one cm^3, then how many grams are in some other volume, like 3 cm^3? You could choose any volume other than one cm^3, such as 2, 3.5, 5, or 10. Choose a volume that makes the calculations easy. For whatever volume you select, record the mass and volume on the graph as a second data point for a particular substance.
4. Connect both points for each substance. You should have four lines on your graph. Clearly label them in a way that a reader can distinguish which line represents which substance.
5. Calculate the slope ($\Delta y/\Delta x$) for each of the lines. Note the slopes on your graph. Show your calculations in the space below.

6. Select four circular objects with diameters and circumferences you can easily measure.
7. You will use the diameter and circumference measurements to make a graph. Title your graph: Circumference vs. Diameter. Label the x-axis as diameter (cm) and the y-axis as circumference (cm).
8. Carefully measure the diameter and circumference for each circle and plot the points. Label the points so a reader will know what objects have been used to plot circumferences and diameters.
9. Draw a best-fit line. If your data points do not approximate a line, you have made an error and will need to check your work.
10. Calculate the slope of the best-fit line. Show your work in the space below.

Concept Development:
11. What is the relationship between the calculated slope and the density for each substance?

12. The number calculated for the slope of the best-fit line should be close to a familiar number associated with circles. What symbol is used to represent this number?

13. In plotting mass vs. volume, you obtained a straight line of different slope for each substance. The slope reflected what intrinsic property for each substance?

14. The equation for a line is often represented as: $y = mx + b$. The circumference for a circle can be represented by: $C = \pi d$. What do the "m" and "π" represent?

Application:
15. Describe an example of slope in "everyday" life.

16. Describe an example of slope (other than any listed above and not for a mathematics class) where slope would be calculated. In the example you list, why is it important that slope be calculated?

Background Reading:
For additional background, read the **Circles** and **Slope** sections of the More Physical Science Reading.

37   Mineral Identification

Related National Science Education Standards:
  Teaching Standard A: inquiry-based
    Science Content: systems, order, and organization
                                        evidence, models, and explanation
                                        constancy, change, and measurement
                                        evolution and equilibrium
                                        form and function
    Science as Inquiry: abilities to do scientific inquiry
                                        understanding about scientific inquiry
    Physical Science: properties of objects and materials
                                        . . . , magnetism
                                        properties and changes of properties of matter
                                        structure and properties of matter
                                        interactions of energy and matter
    Earth and Space Science: properties of Earth materials
                                        geochemical cycles
    Science and Technology: . . . distinguish between natural objects
    Science in Personal and
    Social Perspectives: types of resources
                                        natural resources

Exploration/Awareness:

Minerals, along with other materials that are mined, combine with agricultural products to form the raw materials that societies use. Individual's lives, relationships between countries, and economies are affected by the availability of minerals and their trade.

Minerals can be identified by using their various properties. Variations in specimens or impurities in specimens can be problematic when identifying minerals. For these reasons, it is important to look at a variety of samples. After instruction, use the dichotomous key that follows to identify the minerals. Write the number or letter of the specimen by the name. You may want to read the Properties of Matter and Minerals sections of the More Physical Science Reading before you start this activity.

Dichotomous Key for Mineral Identification

1a. black to gray to silver   go to 2a/2b
1b. not 1a   go to 9a/9b

2a. hardness 2 or less	graphite	_____
2b. not 2a	go to 3a/3b	
3a. elastic sheets	biotite	_____
3b. not 3a	go to 4a/4b	
4a. white streak	go to 5a/5b	
4b. not 4a	go to 6a/6b	
5a. black	anorthite	_____
5b. not a	quartz	_____
6a. "reddish" streak	hematite	_____
6b. not 6a	go to 7a/7b	
7a. black streak	go to 8a/8b	
7b. not 7a	hornblende	_____
8a. magnetic	magnetite	_____
8b. not 8b	galena	_____
9a. brown to yellow to "brassy" yellow	go to 10a/10b	
9b. not 9a	go to 13a/13b	
10a. fingernail will scratch	sulfur	_____
10b. not 10a	go to 11a/11b	
11a. pale yellow streak	sphalerite	_____
11b. not 11a	go to 12a/12b	
12a. black streak	pyrite	_____
12b. not 12a	limonite	_____
13a. "reddish" or reddish brown streak	hematite	_____
13b. not 13a	go to 14a/14b	
14a. scratches glass	go to 15a/15b	
14b. not 14a	go to 19a/19b	
15a. no cleavage	go to 16a/16b	
15b. not 15a	go to 18a/18b	
16a. red-brown	garnet	_____
16b. not 16a	go to 17a/17b	

17a. granular/grainy	olivine	_____
17b. not 17a	quartz	_____
18a. pink-tan	orthoclase	_____
18b. not 18a	albite	_____
19a. elastic sheets	muscovite	_____
19b. not 19a	go to 20a/20b	
20a. white streak	go to 21a/21b	
20b. not 20a	chlorite	_____
21a. fingernail easily scratches	go to 22a/22b	
21b. not 21a	go to 24a/24b	
22a. soapy or greasy feel	talc	_____
22b. not 22a	go to 23a/23b	
23a. earthy luster	kaolinite	_____
23b. not 23a	gypsum	_____
24a. "fizzes" in HCl	calcite	_____
24b. not 24a	go to 25a/25b	
25a. salty taste	halite	_____
25b. not 25a	fluorite	_____

Concept Development:

26. What characteristics or properties are most reliable in helping to identify minerals? Explain.

27. What is the difference between cleavage and fracture?

28. What are some materials that might be used to determine hardness of minerals?

29. What is luster?

Application:
30. Study and practice identifying minerals so that you can identify minerals by sight on the unit exam.

Background Information:
For additional background, read the **Properties of Matter** and **Minerals** sections of the More Physical Science Reading.

38   **More Physical Science Questions and Problems**

1. List two properties that are intrinsic and some substances that can be identified using these two intrinsic properties.

2. List two properties that are accidental. Explain why they are accidental properties and not intrinsic properties.

3. Describe what M/V means without using the terms "density," "mass," or "volume."

4. Based on the density values given in the "Using Graphs" activity, make a general statement that compares the densities of gases, liquids, and solids. What phase or state of matter does not overlap into either of the other phases or states?

5. What does the property of density have to do with whether an object will sink or float?

6. If ice was more dense than water, what might be the consequences for aquatic organisms living in a lake?

7. When water freezes in water pipes and beverage bottles, the pipes and bottles may burst. Explain.

8. When helium is added to a balloon, the balloon will rise in the atmosphere. However, when a person blows air into a balloon it does not rise. Explain why.

9. A volume of 50 cm^3 of a substance is found to have a mass of 965.0 g. Use the density table provided in the "Using Graphs" activity to determine what the substance might be. What other property might be used to identify the substance?

10. A circle has a circumference of 1 meter. What is its diameter?

11. A rock type has a density of 3.6 g/cm^3. If a rock fragment of this type has a mass of 1800 g, what would be its volume?

12. The circumference of the Earth at the equator is about 25000 miles or 40075 km. What is its diameter and radius?

13. Assume the Earth is a perfect sphere. If there are 24 time zones each defined by arcs equal in degree measure and length, what number of degrees defines a time zone at the equator and what distance in miles and km does this arc cover?

14. Earth has a volume of about 1.1 x 10^{27} cm^3 and a mass of about 6.0 x 10^{27} g.
    a. What is the average density of Earth?

    b. Earlier you calculated the density of piece of granite and a piece of basalt. How do they compare to the average density of Earth?

c. What can you infer about the density of the Earth's interior given the average density for Earth and the densities for rocks at the surface?

15. Ethyl alcohol has a density of .791 g/cm^3 and water has a density of 1.000 g/cm^3. When 200 cm^3 of each are combined, the density of the mixture is .91 g/cm^3, but the total volume is not 400 cm^3. Given the information provided here, determine the volume of the mixture.

16. Of solids, liquids, and gases, which:
   a. has a definite shape?

   b. has a definite volume?

   c. is most easily compressed?

17. List five commonly used materials. Identify one property of each material that contributes to its use. Are the properties intrinsic or accidental?

18. _____ is a ratio between the changes in one variable and the changes in another variable on a straight line graph.

(Note: See other Activity-Based Physical Science units for additional reading on selected physical science concepts).

Matter

Physical science is sometimes described as the study of matter and energy. Matter has mass and takes up space. Can you think of something that is not matter? Two pieces of matter can not occupy the same space at the same time.

States or Phases of Matter

The four states of matter include: **solid, liquid, gas,** and **plasma.** Solids have definite volumes and shapes. Liquids have definite volumes but take the shape of their containers. A gas has neither a definite shape nor a definite volume. Plasma is a state of matter that exists at very high temperatures. It is an oversimplification, but it can be useful for some beginning students to think of plasma as being like a very hot gas. Plasma displays some properties of gases but also some properties that differ from those of gases. In plasma, atoms have "lost" their electrons in the sense that the electrons may orbit around more than one atomic nucleus.

Properties of Matter

Properties of matter can be classified in a variety of ways. Properties or characteristics of matter can be classified as properties that help identify a substance or properties that do not help identify a substance. Properties that help identify a substance are called **intrinsic** properties. Properties that do not help identify a substance are called **accidental** or **nonintrinsic** properties. Intrinsic properties include things like density, boiling point, and solubility. Accidental properties include things like mass and volume. Is it possible for a property to be intrinsic in one situation and accidental in another?

Physical properties of matter include properties like freezing and boiling points, odor, color, and hardness. A physical property describes a characteristic of matter that can be observed without changing the chemical composition of the substance. Physical properties of materials are considered when scientists, engineers, and others design products for use. **Chemical properties** are characteristics that depend on the reaction of a substance with other substances. A familiar example of a chemical property is whether a substance will burn.

Note that some references (e. g., NSTA, 1993) may use a different scheme for classifying physical properties of matter such as **intensive** or **extensive** properties. An intensive property has the same value for any subdivision of a system. Examples of intensive properties include pressure, color, odor, texture, hardness, melting point, boiling point, and magnetic character. An extensive property has a value that is the sum of the values of the subdivisions of a system. Examples of extensive properties include mass, length, volume, and surface area.

Scientists have long investigated properties of matter. **Viscosity** is a property that describes the tendency of fluids to resist flow. Low viscosity means the fluid flows easily and high viscosity means the fluid does not flow easily. The **strength** of a material is indicated by its ability to resist changes in shape or breaking. **Electrical properties** of matter are indicated by whether they will conduct electricity easily (**conductors**) or resist the flow of electricity (**insulators**). A similar classification scheme is used to classify materials according to their ability to transmit heat (**thermal properties**). Scientists and technologists have investigated materials that are magnetic and have used this property to produce materials that have benefited society. Other properties of matter include such things as: melting point, boiling point, odor, solubility, elasticity (capable of returning to its original form after deformation), malleability (ability to be rolled or pounded into a thin sheet), ductility (ability to be pulled into a wire), flammability, other types of chemical reactivity (such as the ability to resist rusting), and specific heat. Scientific investigations of matter and their practical applications have resulted in the development of many useful products.

Density

Density is an intrinsic property, i. e., it is a property that helps identify a substance. Also, density is an example of a physical property. Since $D = M/V$, units for density include **g/mL**, **g/cc**, or **g/cm^3**. Density can be thought of as the amount of matter in a certain amount of space. Density is involved in many natural phenomena such as: floating, sinking, interactions of hot and cold air, ocean currents, cloud formation, and many more could be listed. An object or substance floats in a liquid when it is less dense than the liquid. An object will sink in a liquid when it is more dense than the liquid. Earlier, it was noted that physical science is frequently described as the study of matter and energy. Density differences are a major driving force in the distribution of matter and energy on Earth.

Water is most dense at 4° Celsius. During cold weather, the surface layer of fresh water will cool to this temperature and then sink. Warmer water is pushed to the surface and cooled. The convection currents established will result in the entire body of water cooling to 4° C. When the surface layer cools below 4° C, it is less dense than the 4° C water below. Thus, lakes freeze from the top down. What might happen to plants and animals that live in fresh water if this was not the case?

Density measurements are taken in a wide variety of circumstances. Automobile radiator antifreeze/coolant is tested and related to a temperature at which the fluid will freeze (antifreeze in cold weather) or boil (for summer driving). One test that medical technologists can perform on urine is a density test. Under certain physical conditions water has a density of 1.0 g/cm^3. In the case of antifreeze/coolant, a density of 1.0 g/cm^3 would correspond to a freezing point of 0° C or 32° F, the freezing point of water. In the case of urine, there is a healthy range for density that is just a little over the density of water, since urine is mostly water. An illness may cause an excess or deficiency of salts dissolved in urine causing the density to fall outside what is considered normal. The Schlieren effect is the result of a transparent medium such as moving air that has a different density that the rest of the medium. You may have observed this effect when noticing "waves" over a highway. The air over the highway is differentially heated than the air over nearby ground. The interaction of light with the air creates an effect that can be observed.

Buoyancy

Most students have experienced the phenomena where an object submerged in water feels like it weighs less than it does when not submerged. When an object is submerged, the water exerts an upward force opposite the direction of gravitational force. This upward force is called **buoyant force**. Pressure is exerted on an object in a direction perpendicular to its surface. Pressure increases with depth, so there is a greater upward force on the bottom of an object than a downward force on the top of the object. This results in a net force upward, i. e., the buoyant force. If the weight of a submerged object is less than the buoyant force, the object floats. If the weight of a submerged object is greater than the buoyant force, it sinks. What happens when the buoyant force equals the weight?

A completely submerged object displaces a volume of water to equal to its own volume. This is the basis for using the water displacement method for finding volume. In the third century B.C., Archimedes, a Greek scientist, is credited with discovering that an immersed body is buoyed up by a force equal to the weight of the fluid it displaces. This relationship has come to be known as Archimedes' principle.

Changes in Matter

A **physical change** is a change without a change in composition. General examples of a physical change are changes in size, shape, or state (phase). A specific example of a physical change is when H_2O the liquid (water) changes to H_2O the solid (ice). A **chemical change** is a change in which a substance becomes another substance with different properties. It is more than a change in appearance; new substances form. When wood burns or iron rusts, a chemical change has occurred.

Law of Conservation of Mass

Mass is the amount of matter. **Matter** is anything that has mass and takes up space (or has volume). If you did the activities in this unit carefully and took uncertainty into account, the total mass of a system before a change is equal to the total mass after the change. "Doing the activities carefully" would mean that you kept the system closed (nothing was spilled or added) and that no mistakes were made in measuring the masses. In completing two activities in this unit, you should have confirmed the **Law of Conservation of Mass**, i. e., the mass before a change and the mass after a change are equal in ordinary reactions. Any differences should be small and attributed to measurement error.

Scientific inquiry frequently starts with a simple investigation. Insights gained from simple investigations are used to guide investigations in more complex cases. Scientists have repeatedly verified that mass is conserved to the degree that reference is made to the "**law of conservation of mass**." The law of conservation of mass has been arrived at through the process of **induction** or induced, i. e., a general law has been inferred or concluded from looking at specific cases. Scientists support the law of conservation of mass in ordinary reactions because they have not observed cases where it does not hold true, and predictions and inferences based on it have proven to be correct. This is true of other scientific laws, i. e., they are useful in making predictions and there is no evidence that the laws are violated.

Significant Figures

A dictionary definition of "significant figures" or "significant digits" usually makes reference to a number in decimal form. The significant figures or digits are those beginning with the leftmost nonzero digit and extending to the right to include all digits warranted by the accuracy of the measuring devices used to obtain the numbers in the measurement. Whew! For the purposes of this course, the significant figures will be considered the digits used when recording a measurement, i. e., the numbers that can be read plus one more digit (when using a decimal system).

Area and Volume

Area is a term familiar to most students. It can be determined by defining a unit square such as a square one centimeter on each side (1 cm^2) or one meter one each side (1 m^2) and then using this unit square to count how many of them it takes to cover a surface. Counting the number of unit squares is not always easy to do especially when dealing with irregular shapes. Formulas have been derived to use when calculating the areas of regular shapes (see Table 3-7).

In a manner similar to finding area, volume can be determined by defining a unit cube such as one centimeter on each side (1 cm^3) or one meter one each side (1 m^3).

Table 3-7   Area Formulas

Shape or Object	Formula
square	length x length or length^2
rectangle	length x height (or width)
right triangle	½ x base x height
circle	π x radius^2
surface area of a sphere	4 x π x r^2

Then this unit cube can be used to count how many cubes it takes to fill a space occupied by some object. Counting the number of is not always easy to do especially when dealing with irregular shapes. Formulas have been derived to use when calculating the volumes of regular shapes (see Table 3-8).

Table 3-8   Volume Formulas

Shape or Object	Formula
cube	length x length x length or length^3
rectangular solid	length x width x height
sphere	4/3 x π x radius^3
right circular cylinder	π x radius^2 x height
right circular cone	1/3 x π x radius^2 x height

Multiplication is the principle mathematical operation performed in order to calculate area and volume. Multiplication is like repeated addition which is like counting forward. Counting forward is like counting squares or counting cubes. Conversely, division is like repeated subtraction, which is like counting backward. Remember, area was defined as counting the number of squares to cover and volume was defined as counting the number of cubes to fill. These statements are examples of **operational definitions**. Operational definitions refer to the operations or actions that must be carried out to determine numerical values that we give the term being defined. If numbers are not involved, the actions give meaning to the term being defined.

The volume of irregular objects that sink can be measured by using **water displacement**. One way to do this is to use a graduated cylinder and water. A volume of water is placed in the graduated cylinder such that when the object is carefully placed in the graduated cylinder it will be submerged. The volume of the object is the volume of the water plus the object minus the volume of the water initially placed in the graduated cylinder. Also, overflow cans be used to measure the volume of irregular objects. The overflow can is filled with water. A catch can is used to collect the water that flows out of the overflow can after the object is placed carefully in the overflow can.

Capacity is a reference to the amount in a container can hold. This amount is measured in volume units.

Circles

Some essential features of circles include the **circumference** (C) which is the distance around a circle. The **diameter** (d) of a circle is a straight line that passes from a point on the circle through its center to the opposite side. A **radius** (r) is a line segment that runs from the center of a circle to any point on the circle. **Pi** (π) represents the ratio of the circumference of a circle to its diameter. Thus, **C=πd** or **$2\pi r$**. An **arc** represents a portion of the circle or a "curved segment" of the circle. All the way around the a circle represents 360°. Arcs may represent some fractional portion of this 360°.

Slope

Slope is the rate at which the ordinate (y axis) of a point on a line changes with respect to a change in its abscissa (x axis). Slope is sometimes described as the "rise over the run." This is like saying slope = $\Delta y/\Delta x$ or slope is equal to the change in y over the change in x. Related to the x and y values of a point, another way to describe slope is to say that slope is a ratio of y_2-y_1 to x_2-x_1. Slope is one way to determine the relationship between two variables that are graphed with a straight line.

Minerals

A **mineral** is a **naturally occurring, inorganic, homogeneous solid, with a definite chemical composition (within certain limits), and a characteristic crystalline structure**. Minerals and rocks, along with agricultural products, are societies' raw materials. Our present technologic culture is based on the understanding and use of minerals. A relative small number (about 30) of the approximately 3500 known minerals that make up the Earth's crust play a major role in our daily lives. Minerals are studied because of their economic importance and because people have an interest in their surroundings. The importance of minerals in our everyday lives is recognized by a relatively small percentage of the population. Our reliance on minerals can be underscored by tracing human activities for a day and noting the role that minerals play in virtually all human activities.

Color of a mineral can be useful in identification, but it is not always reliable. Streak color is a reliable color. Streak represents the color of a mineral in powdered form achieved by "scratching" the mineral across an unglazed porcelain tile. Cleavage refers to the characteristic break of some minerals along planes of weakness and result in flat surfaces that reflect light. Fracture is any break other than cleavage and can be described using terms like uneven or conchoidal (a smooth curved break like that of thick glass). Luster refers to the quality and quantity of light reflected from the surface of a mineral. Metallic luster can be shiny (bright) metallic or dull metallic. Nonmetallic luster includes: soapy, earthy, greasy, and vitreous (glassy). Hardness is the resistance of a mineral to scratching. Austrian mineralogist F. Mohs chose 10 common minerals to use to compare relative hardness of minerals. The following minerals arranged in order constitute the Mohs hardness scale:

1 – talc	2 – gypsum
3 – calcite	4 – fluorite
5 – apatite	6 – orthoclase
7 – quartz	8 – topaz
9 – corundum	10 – diamond

Hardness is 2 or less if the mineral can be scratched with a fingernail. If a fingernail does not scratch the mineral and the mineral does not scratch glass, the mineral has a hardness between 3 and 5. If the mineral scratches glass, the hardness is greater than 5.

Table 3-9   Mineral Color, Cleavage or Fracture, Hardness, Luster, and Streak

Mineral Name	Color	Cleavage (C) or Fracture (F)	Streak	Luster	Hardness
1. graphite	black	C platy	black	metallic; dull or earthy	1-2
2. sulfur	yellow; yellow with shades green, gray, red	F uneven to conchoidal	yellow	waxy, earthy	1.5-2.5
3. hematite	red, gray	C poor	reddish brown	metallic; dull	5.5-6.5
4. limonite	yellowish brown to dark brown	F uneven	yellowish brown	brilliant to dull; earthy; silky	5-5.5
5. magnetite	black	F uneven	black	metallic	6
6. halite	colorless, white; shades yellow, red, blue, purple	C cubic	white	dull, vitreous; transparent to translucent	2.5

7. fluorite	purple; green, yellow, bluish green	C	white	vitreous	4
8. calcite	colorless to white; variously tinted	C	white	vitreous to earthy	3
9. gypsum	colorless to white or gray; various	C	white	vitreous sometimes pearly or silky	2
10. galena	lead gray	C cubic	lead gray to black	bright metallic	2.5
11. sphalerite	yellow to brown to black; colorless, green, red	C	pale yellow	nonmetallic to resinous to submetallic; brilliant or vitreous	3.5-4
12. pyrite	brassy	F uneven, conchoidal	black; greenish to brownish black	metallic	6-6.5
13. quartz	colorless to white; impurities cause any color	F conchoidal	white	vitreous	7
14. orthoclase	colorless, white, gray, peach; rarely yellow or green	C	white	vitreous	6
15. anorthite	colorless, white, gray, black; green, yellow, peach	C	white	vitreous to pearly	6
16. albite	colorless, white, gray, black; green, yellow, peach	C	white	vitreous to pearly	6
17. muscovite	colorless to yellow,			vitreous to	

(continued) muscovite	brown, green, red	C	white	silky to pearly	2-2.5
18. biotite	dark green to brown to black	C	brownish	vitreous	2.5-3
19. olivine	"green," pale yellow green to olive green	F conchoidal	white	vitreous	6.5-7
20. talc	white, gray, silver-whie, light green	C	white	pearly to greasy	1
21. kaolinite	white	F	white	dull, earthy to pearly	2
22. garnet	dark red, brown, yellow, white, green, black	F conchoidal	white	vitreous to resinous	6.5-7.5
23. hornblende	dark green to black	C	pale green	vitreous to silky	5-6
24. chlorite	shades of green; yellow, white, rose red	C	light green	vitreous to pearly	2-2.5

Scientists

Historical perspectives of science provide concrete examples of the nature of science and illustrations of the profound significance of science to our social and cultural heritage. Many cultures have contributed to our scientific heritage. A few, brief examples illustrating the significance of science follow.

For a long period of time the Sun and other planets were thought to orbit around the Earth. In the 16th century a Polish astronomer, **Nicolaus Copernicus**, is credited with playing a major role in the recognition that the Earth was not the "center of the universe." **Ptolemy**, an Egyptian astronomer working in the 2nd century AD, devised a model that predicted the motions of the Sun, Moon, stars, and planets. The German astronomer, **Johannes Kepler**, used mathematics to provide evidence to support Copernicus' Sun-centered idea. Rather than having the heavenly bodies move in uniform circular motion, he suggested they move in elliptical paths. **Galileo** made observations and discoveries that supported the ideas of Copernicus. His writings were considered controversial for the times and created religious, political, and scientific debates.

Isaac Newton demonstrated that the same laws of motion could be applied on Earth and in space. Newton developed a view of motion based on concepts of force, mass, acceleration, motion, and gravity. His work went beyond physics and astronomy and influenced areas outside the sciences.

Albert Einstein did revolutionary work that related matter and energy and space and time. His work contributed greatly to the human understanding of nature. Included among some of his most famous contributions are the ideas that: the speed of light in a vacuum is constant, nothing travels faster than the speed of light and $E=mc^2$ or energy equals mass times the speed of light squared.

Until the 19th century, most people believed the Earth to be a few thousand years old. **Charles Lyell** used indirect evidences to propose a much greater age of the Earth.

A number of persons have contributed to modern plate tectonic theory. **Alfred Wegener**, early in the 20th century, based on additional evidences reintroduced the idea of moving continents. His failure to provide a mechanism to drive the movement of the continents contributed greatly to this idea's lack of acceptance. It was not until the 1960s after additional supporting evidences were collected that modern plate tectonic theory gained widespread acceptance.

Antoine Lavoisier developed a new field of science that incorporated ideas based on a theory of materials, quantitative methods, and physical laws. Central to his ideas was the conservation of matter. **John Dalton** modernized ancient Greek ideas of atom, molecule, element, and compound. Many scientists, including Lavoisier and Dalton, contributed to the development of chemistry as a science.

The discovery of radioactivity and the structure of the atom and subsequent events, including the use of nuclear energy both as a civil energy source and in weapons, dramatically illustrates the interaction of science, technology, and society. Many scientists were involved. **Marie Curie**, and her husband, **Pierre Curie**, isolated two new radioactive elements. The study of radioactivity provided evidences to scientists resulting in the thought that atoms have a dense nucleus made up of protons and neutrons surrounded by a cloud of electrons. Research by **Ernest Rutherford** and others determined that radioactive uranium splits into a lighter nucleus and a very light helium nucleus. Further work by other scientists showed that when a neutron "bullet" splits uranium the result is two nearly equal parts plus one or two neutrons. **Lisa Meitner** is credited with pointing out that if the fragments add up to less mass than the original mass, a large amount of energy would be released. **Enrico Fermi** showed that the extra neutrons split other nuclei creating a chain reaction capable of producing huge amounts of energy.

Charles Darwin argued for evolution by natural selection. Even though most scientists have accepted Darwin's basic idea, evolution remains a controversial subject among non-scientists. Its rejection by non-scientists is usually not for scientific reasons but because they prefer a religious account of creation.

Louis Pasteur discovered that infectious diseases can be caused by germs. Prior to his discovery, people created many explanations for disease.

The **Industrial Revolution**, like the other examples listed above, illustrates the relationships among science, technology, and society. The Industrial Revolution increased worker productivity and produced other benefits, but also had economic and social consequences that lead to conflict and affected political ideology.

Science Careers

Chemists are employed in a variety of roles. They can serve as researchers that seek new chemical knowledge and practical uses for chemical knowledge. Besides research and development, chemists are employed in teaching, materials and product testing, production, and sales. Specialty areas of chemistry include: analytical, biochemistry, inorganic, organic, and physical. Government agencies and private industries that have quite different missions employ chemists. Environmental testing, developing pharmaceuticals, working with electronic components, and development of synthetic products are just a few examples of areas in which chemists work. A college degree in chemistry is required and many jobs require an advanced degree in chemistry. For more information about chemists and chemistry, contact the American Chemical Society.

Geologists and geophysicists work outdoors, i. e., "in the field," and in laboratories. They study Earth materials and history. Their roles include searching for minerals, oil, gas, and water along with studying rocks, conducting surveys, and measuring other physical aspects of the Earth and some study other planets. Some are employed as teachers. Specialty areas are numerous and include: hydrologist, mineralogist, oceanographer, paleontologist, seismologist, stratigrapher, and vulcanologist. A specialty college degree is required and many jobs require an advanced degree. For more information about geologists and geophysicists, contact the American Geological Institute.

Meteorologists study physical characteristics of the Earth's atmosphere, atmospheric processes, and how they affect the environment. Many are employed as weather forecasters. Other opportunities exist in agriculture, transportation, and air pollution control. Some are employed as teachers. Climatologists analyze past weather records for use in planning and land use. Physical meteorologists research how the atmosphere affects light, sound, and radio transmission. A specialty college degree is required and many jobs require advanced degrees. For more information about meteorologists, contact the American Meteorological Society.

Physicists study the interaction of matter and energy, behavior of matter, and energy transfer and generation. Some are employed as teachers. Most positions require graduate training. For more information about physicists, contact the American Institute of Physics.

Astronomers study the solar system, stars, planets, other astronomical phenomena, and the nature of the universe. Some are employed as teachers. Most positions require graduate degrees.

Science Case Study – Earth's Place in the Universe

The Copernican-Keplerian Revolution that removed the Earth, and humans, from the center of the universe transformed the sense that most people have of their relation to the universe and raised questions of human existence. This case study is a great illustration of how science works and how science, technology, and mathematics are connected. For more than 1500 years people perceived the Earth to be stationary and that all other objects in the sky revolved around the Earth.

In the 2nd century A.D., the Egyptian astronomer Ptolemy devised a model of the universe that predicted motions of the Sun, Moon, stars, and planets. His model incorporated constant motion in perfect circles and even circular motion on smaller circles. Ptolemy's views were consistent with those of Aristotle and other Greek philosophers. In the 16th century Copernicus, a Polish astronomer, suggested observed motions in the heavens could be explained by having Earth rotate once a day and orbit the Sun in a year. His explanation was rejected because it went against common sense and the belief that Earth was at the center of the universe. Protestant and Catholic leaders felt that the Copernican explanation conflicted with their religious beliefs. The German astronomer, Johannes Kepler, used mathematics to show that Copernicus' Sun-centered system worked if, instead of uniform circular motion, predictable motion was uneven but in ellipses.

Galileo used telescopes to make many observations that supported Copernicus' ideas in part by calling traditional views into question. He observed moons of Jupiter, sunspots, moon features such as craters and mountains, and stars previously not visible to the unaided eye. Galileo published arguments for and against the conflicting views of the universe, although favoring the newer view. His writings were controversial scientifically and politically, and upset religious leaders. The work of Kepler, Galileo, and others reflected a change where people had begun to believe the physical world is governed by natural laws. Further, it is possible to discover such laws by making careful measurements, if possible, under controlled laboratory conditions.

accidental properties (examples)

arc

area

average value

axis

buoyancy

buoyant force

capacity

change in y/change in x

characteristics

chemical change

chemical property

circle

circumference

cleavage

color

conservation

conservation of mass

constellations

control variable experiment

coordinate

density (real world examples)

dependent variable

division (repeated subtraction, counting backward)

diameter

elasticity

energy

error (systematic; random)

extensive property

float

fracture

gas

graduated cylinder

graph (constructing, labeling)

hardness

hypothesis

independent variable

inference

intensive property

intrinsic property (examples)

Law of Conservation of Mass

length x width

length x width x height

liquid

luster

manipulated variable

mass

matter

minerals

multiplication (repeated addition, counting forward)

observation

operational definition

physical change

physical property

pi

plasma

prediction

properties of matter (e. g., mass, vol., . . .)

proportional thinking

radius

ratio

recording measurements with uncertainty

rise

run

Schlieren effect

scientific law

scientific notation

significant figures

sink

slope

solid

streak

strength

surface area

theory

uncertainty (as a measurement and percent)

units (length, area, volume, mass, density, . . .)

system (closed; open)

viscosity

volume

water displacement method

x axis (abscissa)

y axis (ordinate)

Part 4   Chemistry

Name(s)

_____

_____

_____

_____

41   Solubility

Related National Science Education Standards:	
Teaching Standard A: inquiry-based	
Science Content:	systems, order, and organization
	evidence, models, and explanation
	constancy, change, and measurement
	evolution and equilibrium
Science as Inquiry:	abilities to do scientific inquiry
	understanding about scientific inquiry
Physical Science:	properties of objects and materials
	properties and changes of properties in matter
	structure and properties of matter
	interactions of energy and matter
Earth and Space Science:	properties of Earth materials
Science and Technology:	understanding about science and technology
History and Nature of Science:	nature of science

Exploration/Awareness:

We are familiar with interactions between substances where the substances mingle with one another and produce a homogeneous (same throughout) mixture. Earlier, you may have completed the activity that asked you to combine salt with water and determine if there was a mass change. Saltwater is an example of a solution. There are lots of examples of solutions.

In this activity you will work with other groups to determine the solubility of salt in water. Each group will determine whether or not the amount of salt assigned to them will completely dissolve in 10 mL (or cm^3) of room temperature water. The collective data will be used to calculate the solubility of salt in water. Your instructor will assign your group number.

1. Use a graduated cylinder to carefully measure 10 mL of room temperature water and pour it into a clean, dry test tube.
2. Use your balance to carefully measure out the amount of salt assigned to your group.
3. Add a small amount of the salt to the water in the test tube, stopper the test tube, and shake it vigorously. Note whether all the salt dissolved.
4. Repeat step 3 until the total amount of salt assigned to your group has been added to the water. Note whether all the salt dissolved.

5. Record your results and the results of others in a data table on the chalkboard or marker board and in the data table that follows.

Group	Assigned Amount of Salt	All Dissolve: Y or N?
1	3.350 g	_____
2	3.400 g	_____
3	3.450 g	_____
4	3.500 g	_____
5	3.550 g	_____
6	3.600 g	_____
7	3.650 g	_____
8	3.700 g	_____

Concept Development:
6. Did your class get consistent data? Explain why or why not.

7. Based on class data, the amount of salt that will dissolve in 10 mL or 10 cm^3 of room temperature water is between:

_____ g   and   _____ g

8. Solubility is defined as the number of grams that will dissolve in 100 cm^3 of water. Your class determined solubility for 10 cm^3 of water. Convert the results from above to g/100 cm^3 of water.

between _____ g/100 cm^3 and _____ g/100 cm^3

9. Suggest some factors that might affect solubility.

10. Is solubility an intrinsic or accidental property? Explain.

11. The solubility of salt (NaCl) in pure water at 20° C is 36 grams per 100 cm^3 of water. How do the class results compare with this value? Explain the results.

Application:

12. In question nine you were asked to suggest some factors that might affect solubility. Design a control variable experiment that would test one factor suggested. What variables would you change? What variables would you keep constant?

13. List two "every day" examples where solubility is a concern.

Background Information:

    For additional background, read the **Solutions** section of the Chemistry Reading.

42   Chromatography

Related National Science Education Standards:
Teaching Standard A: inquiry-based

Science Content:	systems, order, and organization
	evidence, models, and explanation
	constancy, change, and measurement
	evolution and equilibrium
Science as Inquiry:	abilities to do scientific inquiry
	understanding about scientific inquiry
Physical Science:	properties of objects and materials
	position and motion of objects
	properties and changes of properties in matter
	motions and forces
	structure and properties of matter
Earth and Space Science:	properties of Earth materials
History and Nature of Science:	nature of science

Exploration/Awareness:

In this activity, you will use chromatography to separate the constituents of selected water soluble markers.

1. Select three different water soluble markers – one must be a black marker.
2. About 1.5-2.0 cm from the end of a strip of filter paper, draw as thin a line as possible across the filter paper. Let the line dry and then repeat. Do this a total of three or four times.
3. Obtain a large test tube, graduated cylinder, or beaker. Place water in the container at a depth that will allow the end of the suspended filter paper to be in the water but so the marker line will not touch the water.
4. The filter paper needs to be suspended in the container so that it hangs as straight as possible without touching the sides of the container and, as stated above, so that the marker line does not hang in the water. If the marker line touches the water, you will need to start over. To suspend the paper in the container, you can use tape and a pencil or pen.
5. As the water moves up the filter paper, you should get color separation (if any).

Concept Development:

6. Give a reasonable explanation as to why the colors separated, i. e., why would the colors move up the paper at different rates?

7. What measurements could have been made to work at the quantitative or symbolic levels of science understanding with regard to the rates of color movement? In this context, "rate" is being talked about as some measured quantity per time unit.

Application:

8. Compare the process of dissolving or creating a solution with the process of separation by the use of chromatography.

9. Describe a situation where it would be useful to separate substances or identify the constituents of a material.

Background Information:

For additional background, read the **Chromatography** section of the Chemistry Reading.

Name(s) _____

_____

_____

_____

43   Writing Formulas and Naming Compounds

The number of electrons gained, lost, or shared by an atom in forming a compound is called its **oxidation number**. Where an element is located on the periodic table gives an indication of its oxidation number. The sum of the oxidation numbers for atoms in a compound is zero. Some elements have more than one oxidation number. Roman numerals are used to indicate the number for elements with more than one oxidation number.

Figure 4-1   Oxidation Numbers of Selected Elements

1+	2+	3+	1-	2-	3-
copper (I), Cu^{1+}	barium, Ba^{2+}	aluminum, Al^{3+}	bromine, Br^{1-}	oxygen, O^{2-}	nitrogen, N^{3-}
hydrogen, H^{1+}	calcium, Ca^{2+}	chromium, Cr^{3+}	chlorine, Cl^{1-}	sulfur, S^{2-}	phosphorus, P^{3-}
lithium, Li^{1+}	copper (II), Cu^{2+}	iron (III), Fe^{3+}	fluorine, F^{1-}		
potassium, K^{1+}	iron (II), Fe^{2+}		iodine, I^{1-}		
silver, Ag^{1+}	magnesium, Mg^{2+}				
sodium, Na^{1+}	zinc, Zn^{2+}				

Writing formulas for compounds requires following certain procedures. The symbol of the element with the positive oxidation number is written first. **Subscripts**, or numbers written "below," are used to indicate more than one atom or ion. Recall the sum of the oxidation numbers for atoms or ions in a compound totals zero.

Compounds composed of two elements are called **binary compounds**. The names of binary compounds begin with the name of the element with the positive oxidation number. The name of the second element, the one with the negative oxidation number, is changed so that it ends in "ide."

Write formulas for the compounds resulting from a chemical combination of the two elements and write their names:

Figure 4-2   Binary Compounds

Elements	Formula	Name
1. sodium and chlorine		
2. hydrogen and sulfur		
3. potassium and iodine		
4. copper (I) and sulfur		
5. aluminum and oxygen		

Polyatomic ions are groups of atoms, that act together as one ion. Writing formulas containing polyatomic ions follow procedures similar to those listed earlier. When naming compounds with a polyatomic ion having a negative charge, do not change the ending to "ide."

Figure 4-3   Charges of Selected Polyatomic Ions

1+	1-	2-	3-
ammonium, NH_4^{1+}	acetate, $C_2H_3O_2^{1-}$	carbonate, CO_3^{2-}	phosphate, PO_4^{3-}
	chlorate, ClO_3^{1-}	sulfate, SO_4^{2-}	
	hydroxide, OH^{1-}		
	nitrate, NO_3^{1-}		

Write formulas for the compounds resulting from a chemical combination involving one or more polyatomic ions write their names:

Figure 4-4   Compounds with Polyatomic Ions

Element and/or Polyatomic Ion(s)	Formula	Name
6. hydrogen and hydroxide		
7. ammonium and phosphate		
8. silver and nitrate		
9. calcium and phosphate		
10. aluminum and hydroxide		

11. What kind of atoms and how many of each are illustrated by the following formulas?

a. H_2O

b. H_2O_2

c. NH_4NO_3

d. $Al_2(SO_4)_3$

44   **Chemical Reactions**

Related National Science Education Standards:
  Teaching Standard A: inquiry-based
    Science Content: systems, order, and organization
                                        evidence, models, and explanation
                                        constancy, change, and measurement
                                        evolution and equilibrium
    Science as Inquiry: abilities to do scientific inquiry
                                          understanding about scientific inquiry
    Physical Science: properties of objects and materials
                                          . . . , heat, . . .
                                          properties and changes of properties in matter
                                          motions and forces
                                          transfer of energy
                                          structure and properties of matter
                                        chemical reactions
                                          interactions of energy and matter
    History and Nature of
    Science: science as a human endeavor
                                          nature of science

Exploration/Awareness:

    In this activity you will have the opportunity to observe different types of chemical reactions. All proper safety procedures must be followed. Dispose of the chemicals as directed by your instructor.

1. Decomposition reaction:
   A decomposition reaction occurs when a substance breaks down into two or more other substances.

   Example 1: Over time hydrogen peroxide decomposes to form oxygen and water.

   $$2 \, H_2O_2 \rightarrow O_2 + 2 \, H_2O$$

   Example 2: Carbon dioxide, water, and sodium carbonate are the products when sodium bicarbonate (baking soda) decomposes.

   $$2 \, NaHCO_3 \rightarrow CO_2 + H_2O + Na_2CO_3$$

2. Synthesis reaction: A synthesis reaction is when two or more substances combine to form one compound.

   Example 1: The reactants carbon and oxygen combine to form the product, carbon dioxide.

   $$C + O_2 \rightarrow CO_2$$

   Example 2: The "rusting" of iron occurs when iron combines with oxygen.

   $$4\ Fe + 3\ O_2 \rightarrow 2\ Fe_2O_3$$

3. Single replacement (or displacement) reaction: A single replacement or displacement reaction is where one element replaces another element in a compound.

   Example 1: Silver forms on copper wire when copper wire is placed in silver nitrate solution.

   $$2\ AgNO_3 + Cu \rightarrow 2\ Ag + Cu(NO_3)_2$$

4. Double replacement (or displacement) reaction: A double replacement or displacement reaction occurs when the positive part of one compound combines with the negative part of another compound.

   Example 1: When hydrochloric acid reacts with sodium hydroxide, the hydrogen ion combines with the hydroxide ion and the sodium ion combines with the chloride ion.

   $$HCl + NaOH \rightarrow HOH + NaCl$$

5. Exothermic reaction:
   a. Measure about 100 mL of water in a graduated cylinder. Record the exact amount in the Exothermic Reaction table. Pour the water into a foam (like Styrofoam®) cup.
   b. Determine the temperature of the water and record it in Table 4-1.
   c. Weigh about 3 grams of sodium hydroxide or about 5 grams of calcium chloride. Record the chemical used and the exact mass in the table.
   d. Add the sodium hydroxide or the calcium chloride to the water in the foam cup. Stir the solution with a glass stirring rod. Read and then record the highest temperature.
   e. Dispose of the contents and clean the equipment as directed by your instructor.

Table 4-1   Exothermic Reaction

Volume of water used	_____
Temperature of water before sodium hydroxide or calcium chloride was added	_____
Mass of sodium hydroxide or calcium chloride used	_____
Temperature of water after sodium hydroxide or calcium chloride was added	_____
Temperature change	_____

6. Endothermic reaction:
   a. Measure about 100 mL of water in a graduated cylinder. Record the exact amount in the Endothermic Reaction table. Pour the water into a foam (like Styrofoam®) cup.
   b. Determine the temperature of the water and record it in Table 4-2.
   c. Weigh about 5 grams of ammonium nitrate. Record the exact mass in the table.
   d. Add the ammonium nitrate to the water in the foam cup. Stir the solution with a glass stirring rod. Read and then record the lowest temperature.
   e. Dispose of the contents and clean the equipment as directed by your instructor.

Table 4-2   Endothermic Reaction

Volume of water used	_____
Temperature of water before ammonium was added	_____
Mass of ammonium nitrate used	_____

Temperature of water after ammonium nitrate was added	_____
Temperature change	_____

Concept Development:

7. With the help of your instructor, identify the reactants in each of the above different types of reactions.

8. With the help of your instructor, identify the products in each of the above different types of reactions.

9. Which reactions produced a change in temperature? Was heat added or removed?

Application:

10. Pour 30 mL of hydrogen peroxide and add 15 mL of water into a clear, transparent plastic cup. Use a thermometer to take the temperature, then record the temperature in a data table as "time 0."

11. Measure 5 mL of yeast. Have one lab partner ready to read the thermometer and a second lab partner ready to call out the time every 10 seconds.

12. Pour the yeast into the cup with the hydrogen peroxide and water. Swirl the contents and every 10 seconds read the temperature. Record the time and temperature data in the data table.

13. Make a graph of the data. What does the graph illustrate about the reaction?

14. Pour 30 mL of vinegar into a clear, transparent plastic cup. Use a thermometer to take the temperature, then record the temperature in a data table as "time 0."

15. Measure 5 mL of baking soda. Be prepared to record time and temperature data every 3 seconds. Taking data every 3 seconds, record the time and temperature data in the data table.

16. Make a graph of the data. What does the graph illustrate about the reaction?

Background Information:
      For additional background, read the **Chemical Reactions** section of the Chemistry Reading.

45   Introduction to Acids, Bases, and Indicators

Related National Science Education Standards:	
Teaching Standard A: inquiry-based	
Science Content:	systems, order, and organization
	evidence, models, and explanation
	constancy, change, and measurement
	evolution and equilibrium
Science as Inquiry:	abilities to do scientific inquiry
	understanding about scientific inquiry
Physical Science:	properties of objects and materials
	properties and changes of properties in matter
	structure and properties of matter
Earth and Space Science:	properties of Earth materials
History and Nature of	
Science:	nature of science

Exploration/Awareness:

Common acids and bases play important roles in daily life. In this introduction to acids, bases, and indicators, you will first test some household substances with litmus paper, pHydrion paper, and red cabbage extract. Follow proper safety procedures!

1. Pour a small amount of the substance to be tested into a test tube or beaker and then immerse a part of the litmus paper or pHydrion paper in the substance.
2. Observe and record the result in Table 4-3.
3. Based on the reaction with litmus paper, classify the substance as acid, base, or neutral. Again, record this information in the table.
4. Finally, pour an equal amount of red cabbage extract in the test tube and note the color change (if any) when the substance is mixed with red cabbage extract. Record the color in the table.

Concept Development:

5. What substances were acids? What substances were bases? What substances were neutral?

6. As an indicator, red cabbage extract is what color in a base?

Table 4-3   Acids, Bases, and Indicators

Substance	Effect upon red litmus	Effect upon blue litmus	Acid, Base, or Neutral?	Color when combined with red cabbage	pHydrion number
1					
2					
3					
4					
5					
6					
7					
8					

7. As an indicator, red cabbage extract is what color in an acid?

Application:
    A number of substances can be used as indicators. You may have noticed how the color of tea changes when lemon juice is added. Consider doing a control variable experiment to identify other common substances that can be used as indicators.

8. Use the pHydrion paper results and combine them with the red cabbage extract results to develop a number scale for the red cabbage indicator colors.

9. Identify a circumstance where testing to determine acid or base levels might be done.

Background Information:
      For additional background, read the **Acids**, **Bases**, and **Salts** and **Indicators** sections of the Chemistry Reading.

46   **Acid-Base Reactions – Practice 1**

Note: The following are examples of chemical changes and the Law of Conservation of Mass.

1. $\overset{1+ \ 1-}{NaOH}$ + $\overset{1+ \ 1-}{HCl}$ →

2. $\overset{1+ \ 1-}{KOH}$ + $\overset{1+ \ 1-}{HCl}$ →

3. $\overset{1+ \ 1-}{HNO_3}$ + $\overset{2+ \ 1-}{CaOH}$ →

4. $\overset{3+ \ 1-}{AlOH}$ + $\overset{1+ \ 1-}{HNO_3}$ →

5. $\overset{1+ \ 2-}{HSO_4}$ + $\overset{2+ \ 1-}{MgOH}$ →

6. $\overset{1+ \ 1-}{NaOH}$ + $\overset{1+ \ 2-}{HSO_4}$ →

7. $\overset{2+ \ 1-}{MgOH}$ + $\overset{1+ \ 2-}{HCO_3}$ →

8.
$$\overset{1+ \ 2-}{HCO_3} \quad + \quad \overset{1+ \ 1-}{KOH} \quad \rightarrow$$

9.
$$\overset{2+ \ 1-}{CaOH} \quad + \quad \overset{1+ \ 3-}{HPO_4} \quad \rightarrow$$

10.
$$\overset{1+ \ 3-}{HPO_4} \quad + \quad \overset{3+ \ 1-}{AlOH} \quad \rightarrow$$

11. What pattern or relationship do you notice between the coefficients and the subscripts?

Name(s) _____

_____

_____

_____

47   **Acid-Base Reactions – Practice 2**

Note: The following are examples of chemical changes and the Law of Conservation of Mass.

1. $\overset{1+\ 1-}{\text{KOH}}$ + $\overset{1+\ 1-}{\text{HCl}}$ →

2. $\overset{2+\ 1-}{\text{MgOH}}$ + $\overset{1+\ 1-}{\text{HCl}}$ →

3. $\overset{1+\ 3-}{\text{HPO}_4}$ + $\overset{2+\ 1-}{\text{CaOH}}$ →

4. $\overset{1+\ 2-}{\text{HSO}_4}$ + $\overset{3+\ 1-}{\text{AlOH}}$ →

5. $\overset{1+\ 2-}{\text{HSO}_4}$ + $\overset{2+\ 1-}{\text{CaOH}}$ →

6. $\overset{3+\ 1-}{\text{AlOH}}$ + $\overset{1+\ 3-}{\text{HPO}_4}$ →

7. $\overset{2+\ 1-}{\text{BaOH}}$ + $\overset{1+\ 2-}{\text{HCO}_3}$ →

8. $\overset{1+\ 2-}{HCO_3}\ +\ \overset{1+\ 1-}{NaOH}\ \rightarrow$

9. $\overset{2+\ 1-}{MgOH}\ +\ \overset{1+\ 3-}{HPO_4}\ \rightarrow$

10. $\overset{1+\ 3-}{HXO_4}\ +\ \overset{3+\ 1-}{ZOH}\ \rightarrow$

48   Combining Acids and Bases

Different volumes of the same acid and base were combined in six trials. The following data were obtained:

Trial No.	Acid Used	Base Used	Effect of Resulting Solution on: Red Litmus	Blue Litmus
1	56 mL	28 mL	blue	blue
2	50 mL	10 mL	red	blue
3	30 mL	30 mL	blue	blue
4	48 mL	9 mL	red	red
5	70 mL	14 mL	red	blue
6	19 mL	3 mL	red	red

Using these data and knowledge of red and blue litmus paper as indicators, determine whether the statements are true (**T**) or false (**F**). If the data and knowledge of litmus paper as indicators together do not allow a true or false decision to be made, label the statement "**U**" for unknown.

_____ 1. The acid and base have equal concentrations.

_____ 2. A base and an acid can react to neutralize one another.

_____ 3. With this acid and base, when the volume of acid used is greater than the volume of base used, resulting solutions will be acid or neutral.

_____ 4. One milliliter of the base neutralizes 5 mL of the acid.

_____ 5. The resulting trial 2 solution would have no effect on phenolphthalein.

_____ 6. The resulting trial 6 solution has a greater amount of hydrogen ions as acid than the solution resulting from trial 4.

_____ 7. When acid reacts with base the reaction yields salt plus water.

_____ 8. The strength of the acid is .2 times the strength of the base.

49　　　**BTB Testing**

Related National Science Education Standards: 　Teaching Standard A: inquiry-based 　　Science Content:　　　　　systems, order, and organization 　　　　　　　　　　　　　　　evidence, models, and explanation 　　　　　　　　　　　　　　　constancy, change, and measurement 　　　　　　　　　　　　　　　evolution and equilibrium 　　Science as Inquiry:　　　　abilities to do scientific inquiry 　　　　　　　　　　　　　　　understanding about scientific inquiry 　　Physical Science:　　　　　properties of objects and materials 　　　　　　　　　　　　　　　properties and changes of properties in matter 　　　　　　　　　　　　　　　structure and properties of matter 　　Earth and Space Science:　properties of Earth materials 　　Science in Personal and 　　Social Perspectives:　　　　personal health 　　History and Nature of 　　Science:　　　　　　　　　nature of science

Exploration/Awareness:

In this activity two people in your group will "bubble" their breath through bromothymol blue (BTB) indicator solution (0.04%) and analyze the results.

1. Place 60 mL of water in a plastic cup. Add 10 drops of BTB solution. Use a straw to mix the solution.
2. Two different people in your group need to "bubble" their breath through a clean straw and into separate cups with the BTB solution. In seconds, time how long it takes for the BTB solution to change from "blue" to "yellow" (or "yellowish-green").
3. Record each person's time to cause the BTB indicator to change color.
4. Using 0.1 M NaOH, see how many drops it takes to return the BTB solution to its original blue color. Record this information for each person.
5. Construct graphs that show person vs. time in seconds and person vs. number of drops for at least 6 people.

Concept Development:

Your exhaled breath contains CO_2 and the cup originally contained water and BTB indicator. The reaction written simply is:

$$\overset{4+ \ 2-}{CO_2} + \overset{1+ \ 2-}{H_2O} \rightarrow \overset{1+ \ 4+ \ 2-}{H_2CO_3}$$

6. As written, is this a balanced chemical reaction? Why or why not?

7. What indicator color indicates base and what color indicates acid?

8. How many drops of base were needed to return your group members' "breath water" to blue? Whether the numbers of drops were equal or unequal, explain why.

9. What variables besides time should be considered when determining how long it takes a person's breath to turn the solution acidic?

10. Why is controlling variables an important part of this or any other experiment?

Application:
11. List a "real-world" example of how conducting a control variable experiment might allow you to make a better decision.

12. BTB is sometimes purchased in pet stores to be used by people that have fish for pets. What are they testing with BTB?

Background Information:
    For additional background, read the **Acids**, **Bases**, **Salts**, and **Indicators** sections of the Chemistry Reading.

50   **Comparing the Acid in Two Types of Soda/Pop**

Related National Science Education Standards:	
Teaching Standard A: inquiry-based	
Science Content:	systems, order, and organization
	evidence, models, and explanation
	constancy, change, and measurement
	evolution and equilibrium
Science as Inquiry:	abilities to do scientific inquiry
	understanding about scientific inquiry
Physical Science:	properties of objects and materials
	properties and changes of properties in matter
	structure and properties of matter
Earth and Space Science:	properties of Earth materials
Science in Personal and	
Social Perspectives:	personal health
History and Nature of	
Science:	nature of science

Exploration/Awareness:

In this activity you will compare the amount of acid in different types of soda/pop. As a group, you only need to determine the amount of base needed to neutralize the acid in the soda/pop assigned to you and then share your data with other lab groups. Follow proper safety procedures. YOU MUST WEAR SAFETY GOGGLES WHEN TITRATING THE SODA/POP WITH THE BASE!

1. Predict what type of soda/pop you think will have the most acid.

2. Check and note the 0.1 M NaOH level in the buret. You may or may not need to add more 0.1M NaOH solution depending on the level. It is important to remember the starting point.
3. Use a graduated cylinder to carefully measure 20 mL of soda/pop and pour it into a clean beaker.
4. Add four drops of phenolphthalein (indicator) to the soda/pop in the beaker.
5. Place a piece of white paper underneath the beaker.
6. Slowly add the 0.1 M NaOH solution, constantly swirling the soda/pop. When the soda/pop appears to retain a "pinkish" color, start adding the 0.1 M NaOH drop by drop. Again, you must swirl the solution. When the soda/pop reatains a "pinkish" color, stop! Note: cola colored drinks will be "reddish-brown."

7. Record the number of milliliters of 0.1 M NaOH used by subtracting the "end point" from the starting point in the buret. The more NaOH used, the more acidic the soda/pop. Record your results and the results of others in the Soda/Pop Comparison table.

Table 4-4   Soda/Pop Comparison

Type of Soda/Pop	Amount of 0.1 M NaOH to Neutralize

Concept Development:

8. Which type of soda/pop had the most acid in it? Is it the one you predicted?

9. Are the results consistent for each of the different brands of soda/pop? If not, what could you do to improve the results of this experiment?

10. What color does phenolphthalein turn in an acid? What color does it turn in a base?

11. Use $M_1V_1 = M_2V_2$ to calculate the molarity of the acid in both of the sodas/pops.

Application:

12. What effect could drinking soda/pop have on your teeth? What other factors, besides effect on your teeth, might influence decisions about which brand of pop to buy?

13. If you are searching for control variable experiment ideas, what other substances might you compare to see which contains the most acid?

Background Information:

For additional background, read the **Acids**, **Bases**, **Salts**, and **Indicators** sections of the Chemistry Reading.

51   Comparing the Neutralization Power of Antacids

Related National Science Education Standards:
  Teaching Standard A: inquiry-based
    Science Content: systems, order, and organization
                                 evidence, models, and explanation
                                 constancy, change, and measurement
                                 evolution and equilibrium
    Science as Inquiry: abilities to do scientific inquiry
                                 understanding about scientific inquiry
    Physical Science: properties of objects and materials
                                 properties and changes of properties in matter
                                 structure and properties of matter
    Earth and Space Science: properties of Earth materials
    Science in Personal and
    Social Perspectives: personal health
    History and Nature of
    Science: nature of science

Exploration/Awareness:

In this activity you will compare antacid tablets by determining the amount of acid each will neutralize. As a group, you only need to determine the amount of acid needed to neutralize the antacid tablet assigned to you and then share the data with other lab groups. Which type of antacid tablet do you think will neutralize the most acid? Follow proper safety procedures. YOU MUST WEAR GOGGLES WHEN TITRATING THE ANTACID WITH THE ACID!

1. Check and note the 0.1 M HCl level in the buret. You may or may not need to add more 0.1 M HCl solution depending on the level. It is important to remember the starting point.
2. Use a mortar and pestle to powder the tablet. Place the powdered tablet in a clean, dry beaker and use a graduated cylinder to add 250 mL water.
3. Add six drops of phenolphthalein (indicator) to the dissolved antacid tablet in the beaker.
4. Place a piece of white paper underneath the beaker. The dissolved tablet in solution with the addition of phenolphthalein needs to turn pink for you to continue. If it did, continue with the neutralization reaction. If not, you will record: "No pink color after the addition of phenolphthalein" and share this information with other groups. Do not continue with the neutralization reaction.
5. Slowly add the 0.1 M HCl solution, constantly swirling the solution. When the solution appears to have permanently lost its "pinkish" color, start adding the 0.1 M

HCl drop by drop. Again, you must swirl the solution. When the solution loses its "pinkish" color, and the pink color does not come back, stop.

6. Record the number of milliliters of 0.1 M HCl used by subtracting the "end point" from the starting point in the buret. The more HCl used, the more acid the antacid neutralized. Record your results and the results of others in the Antacid Comparison Data Table.

Table 4-5   Antacid Comparison

Antacid	Amount of 0.1 M HCl to Neutralize

Concept Development:

7. Which type of antacid neutralized the most acid? Is this the one you predicted?

8. Are the results consistent for each of the different brands of antacid? If not, what could be done to improve the results of this experiment?

9. Use $M_1V_1 = M_2V_2$ to calculate the molarity of one of the antacid solutions, i. e., one of the solutions produced by adding a crushed antacid tablet to water.

Application:

10. Can you unequivocally state the antacid tablet that neutralized the most acid in this activity is the one that a person should use to solve their acid indigestion problems? Why or why not?

11. If you are searching for control variable experiment ideas, what other substances might you compare to see which is most basic?

Background Information:

For additional background, read the **Acids**, **Bases**, and **Salts** sections of the Chemistry Reading.

52   **Designing a Control Variable Experiment**

Related National Science Education Standards:	
Teaching Standard A: inquiry-based	
Science Content:	systems, order, and organization
	evidence, models, and explanation
	constancy, change, and measurement
	evolution and equilibrium
Science as Inquiry:	abilities to do scientific inquiry
	understanding about scientific inquiry
Physical Science:	properties of objects and materials
	properties and changes of properties in matter
	structure and properties of matter
Earth and Space Science:	properties of Earth materials
Science in Personal and Social Perspectives:	personal health
History and Nature of Science:	nature of science

Introduction:

When opening a bottle of aspirin you note a smell somewhat similar to the smell of vinegar (acetic acid, CH_3COOH – note this acid does not begin with an "H" in its formula!). You recall that your instructor mentioned that aspirin is composed of acetyl-salicylic acid ($CH_3COOC_6H_4COOH$). You wonder if the aspirin changes to vinegar. This might be hard to determine with the equipment that you have to work with. Instead you decide to test whether there is a difference between "new" aspirin (should have lots of acetylsalicylic acid and little acetic acid) and "old" aspirin (should have more acetic acid if your hypothesis is correct) by using a neutralization reaction.

YOUR TASK: Design and write up a control variable experiment that would compare old and new aspirin using neutralization reactions. What variables would be kept the same? What variables would be allowed to change?

53   **Temperature, Heat Energy, and Evaporation**

Related National Science Education Standards:	
Teaching Standard A: inquiry-based	
Science Content:	systems, order, and organization
	evidence, models, and explanation
	constancy, change, and measurement
	evolution and equilibrium
	form and function
Science as Inquiry:	abilities to do scientific inquiry
	understanding about scientific inquiry
Physical Science:	properties of objects and materials
	position and motion of objects
	. . . , heat, . . .
	properties and changes of properties in matter
	transfer of energy
	structure and properties of matter
	conservation of energy . . .
	interactions of energy and matter
Earth and Space Science:	properties of Earth materials
Science and Technology:	understanding about science and technology
Science in Personal and Social Perspectives:	science and technology in society
History and Nature of Science:	nature of science

Exploration/Awareness:

     In this activity you will explore the relationship between evaporation, heat, temperature.

1. Water and rubbing alcohol left in a room for an extended time should end up being at "room temperature."
2. Use a thermometer to establish what room temperature is and record it in the Room Temperature: Air – Water – Alcohol table.
3. Use relative terms like: "cool," "warm," "hot," "colder," "tepid," . . . to describe how the air temperature feels to one member of your group when that person holds their finger in the air. Record information in the table.
4. Have the same group member dip a finger into the room temperature water. Hold this finger in the air and, again, use relative terms to describe how the air temperature feels. Record the information in the table.

5. Have the same group member dip a finger into the room temperature alcohol. Hold this finger in the air and, again, use relative terms to describe how the air temperature feels. Record the information in the data table.

Table 4-6   Room Temperature: Air – Water – Alcohol

Test	Air	Water	Alcohol
actual temperature			
how temperature feels to finger held in air			

6. Use a twist tie or rubber band to attach a cotton ball or two cotton balls to each bulb of two thermometers. If the thermometers have been in the room and unused, they should record room temperature. Record room temperature (air temperature) and the starting temperatures of the water and alcohol (which should be at room temperature) in Table 4-7.
7. One person in your group should test room temperature water and another person should test room temperature rubbing (isopropyl) alcohol. Dip the cotton-covered thermometer bulb of one thermometer in the water at the same time the cotton-covered thermometer bulb of the other thermometer is being dipped into the alcohol. Squeeze the excess liquid back into the container.
8. Place the thermometers upside down in a beaker large enough to support the thermometers without tipping over. After five minutes, record the temperatures in the Table 4-7.
9. Determine the temperature change and record this in the table.

Table 4-7   Temperature and Evaporation

Temperature (°C)	Cotton with Water	Cotton with Alcohol
starting temperature		
temperature after 5 minutes		
change in temperature		

Concept Development:

10. Does the evaporation of water remove heat or add heat? What is your evidence?

11. Does the evaporation of alcohol remove heat or add heat? What is your evidence?

12. Compare the data from the evaporation of alcohol with the data from the evaporation of water. Based on these data, what can you infer?

13. Did the thermometers work when they were upside down? Provide a good reason why the thermometers were placed upside down (other than the instructions said to do this).

14. What is the evidence that the evaporation process removes heat from you or your group member's finger?

15. How is evaporation a cooling process, i. e., how is heat removed?

Application:

16. How does perspiring or sweating help cool the body?

17. Name two appliances where evaporation is used to cool food.

18. What form of technology uses evaporation to cool homes?

19. Alcohol baths are sometimes prescribed to reduce high fevers. Why?

20. Does it take more energy for molecules of water and alcohol to exist as liquids or to exist as gases? Explain.

21. What can be done to speed up the evaporation process?

Background Information:

For additional background, read the **Kinetic Theory of Matter** and **Heat** sections of the Chemistry Reading.

54 **Phase Changes of Water**

Related National Science Education Standards:
Teaching Standard A: inquiry-based
Science Content: systems, order, and organization
evidence, models, and explanation
constancy, change, and measurement
evolution and equilibrium
Science as Inquiry: abilities to do scientific inquiry
understanding about scientific inquiry
Physical Science: properties of objects and materials
heat, . . .
properties and changes of properties in matter
transfer of energy
structure and properties of matter
interactions of energy and matter
Earth and Space Science: properties of Earth materials

Exploration/Awareness:

Water, or H_2O, is a unique substance because it exists as a solid, liquid, and gas within the range of temperatures found on Earth. Temperature data were collected as ice was heated to change first to water and then to steam. An assumption made is that the heat source provided energy at a constant rate so that the time heated is proportional to the energy transferred to the H_2O.

Table 4-8   Heating Data

Time (minutes; α energy)	Temperature (° C)	Time (minutes; α energy)	Temperature (° C)
0	-10	21	80
2	0	24	100
4	0	28	100
5	0	31	100
8	15	34	100
12	35	36	110
14	45	37	115
17	60	38	120

1. Make a line graph with temperature as the dependent or responding variable and time as the independent or manipulated variable.

Concept Development:

2. Is there a direct relationship between the time heated and the temperature? Explain.

3. At what temperature does ice melt? What region of the graph does this represent?

4. At what temperature does water boil? What region of the graph does this represent?

5. If energy is being added constantly, why doesn't the temperature increase proportionately to the time?

6. Use C_p for ice as 0.492 cal/g °C, C_p for water as 1.000 cal/g °C, and C_p for steam as 0.485 cal/g °C along with the equation that follows to calculate the total amount of energy required to change 20 grams of water from −10 °C to 120 °C. Show your work!

$$\text{for ice} \qquad \text{for water} \qquad \text{for steam} \qquad s \longleftrightarrow l \qquad l \longleftrightarrow g$$
$$\Delta H = (\Delta T \times m \times C_p) + (\Delta T \times m \times C_p) + (\Delta T \times m \times C_p) + (80 \text{ cal/g} \times m) + (540 \text{ cal/g} \times m)$$

Application:

7. Earlier, you identified the melting and boiling point temperatures. What is the freezing point temperature for water?

8. Sketch the shape of a graph if you were cooling steam to ice.

9. Give an example where water vapor is at the same temperature as water (liquid).

Background Information:

      For additional background, read the **Phase Changes** section of the Chemistry Reading.

55   **Ice Cream Chemistry**

Related National Science Education Standards:
Teaching Standard A: inquiry-based
Science Content: systems, order, and organization
evidence, models, and explanation
constancy, change, and measurement
evolution and equilibrium
form and function
Science as Inquiry: abilities to do scientific inquiry
understanding about scientific inquiry
Physical Science: properties of objects and materials
. . . , heat, . . .
properties and changes of properties in matter
motions and forces
transfer of energy
structure and properties of matter
interactions of energy and matter
Science and Technology: abilities of technological design
understanding about science and technology
Science in Personal and
Social Perspectives: personal health
History and Nature of
Science: science as a human endeavor
nature of science

Exploration/Awareness:

In this activity you will have the opportunity to make homemade ice cream. It is important to follow proper safety precautions if you plan to eat the ice cream. Follow all necessary safety requirements. Check with your instructor to see if eating the ice cream is allowed.

Commercial ice cream makers can be used (using one powered by electricity is much easier than one that needs to be powered by humans!), along with your favorite ice cream recipe, as an option that might be easiest. However, if an ice cream maker is not available and you do not have a favorite recipe, try the following and make adjustments as experience dictates. Once again, if eating ice cream is allowed, make sure you follow proper safety procedures and good health practices!

1. Make sure all materials are clean. Follow appropriate safety procedures and health practices.
2. You need a large, clean, resealable plastic bag and a small, clean, resealable plastic bag. Another option is to use different sized coffee cans.

3. In a large bowl, mix 1 cup of cream, ½ cup of milk, 2 eggs, ¼ teaspoon of vanilla, and ½ cup sugar. Add ¼ cup of chocolate syrup if you want chocolate ice cream.
4. After mixing, put the ingredients in the small, clean, resealable plastic bag.
5. Securely seal the bag.
6. Put the small bag into the large bag. Fill in the space around the smaller bag with ice and add some salt.
7. Securely seal the large bag.
8. While outside, toss the bags until the ice melts. Wear gloves to keep your hands from getting cold.
9. Fill with ice and add salt again and repeat until ice cream forms on the sides of the small bag.
10. Use a thermometer to measure the temperature of the ice and salt water.

Concept Development:
11. What is the purpose of the ice?

12. What is the purpose of the salt?

13. What is the freezing point of water? Was the temperature of the slush higher than, equal to, or less than the freezing point of water?

14. What phase changes took place in this activity?

15. Does the ice melting in the bag mean the temperature is going up?

Application:
16. What could be done to improve the technology?

17. What could be done to improve the product?

18. Identify another situation where salt is used to change the freezing point of water.

Background Information:
    For additional background, read the **Heat and Temperature** and **Phase Changes** sections of the Chemistry Reading.

Name(s) _____

_____

_____

_____

56   Water Quality Testing

Related National Science Education Standards:
  Teaching Standard A: inquiry-based
    Science Content: systems, order, and organization
                                 evidence, models, and explanation
                                 constancy, change, and measurement
                                 evolution and equilibrium
                                 form and function
    Science as Inquiry: abilities to do scientific inquiry
                                 understanding about scientific inquiry
    Physical Science: properties of objects and materials
                                 properties and changes of properties in matter
                                 structure and properties of matter
                                 interactions of energy and matter
    Life Science: . . . environments
                                 . . . ecosystems
                                 matter, energy, . . . in living systems
    Earth and Space Science: properties of earth materials
                                 changes in earth and sky
                                 geochemical cycles
    Science and Technology: understanding about science and technology
    Science in Personal and
    Social Perspectives: personal health
                                 types of resources
                                 changes in environments
                                 science and technology in local challenges
                                 natural hazards
                                 risks and benefits
                                 science and technology in society
                                 environmental quality
                                 natural and human-induced hazards
    History and Nature of
    Science: science as a human endeavor
                                 nature of science

Exploration/Awareness:

In this activity you will have the opportunity to determine level of water quality using test kits that measure nitrates, hardness, and dissolved oxygen. All proper safety procedures must be followed. Dispose of chemicals as directed by your instructor. Your instructor will provide directions that included in the kits to complete this activity.

Concept Development:

1. List some sources of nitrates.

2. How are nitrates useful? How are they harmful?

3. What is dissolved oxygen? How is it measured?

4. What factors may affect levels of dissolved oxygen?

5. What is water hardness?

6. What factors affect levels of water hardness?

Application:

7. Test a water sample designated by your instructor and report results.

Background Information:

      For additional background, read the **Water** and **Water Pollution** sections of the Chemistry Reading.

57   **Chemistry-Related Hobby Report**

Significant numbers of people engage in chemistry-related activities outside their time at work. They pursue these activities as hobbies out of curiosity, for the enjoyment it brings, and/or for the benefits it brings to their lives.

What follows is a list of some hobbies or activities where possession of some chemistry knowledge and/or skills would be useful. Other possibilities exist.

candle making	ceramics
cooking	furniture refinishing
gardening	

Select one of the possibilities from the list or come up with a chemistry-related hobby or activity of your own (get your instructor's approval on your idea). Write a 1 ½ to 2 page, double-spaced report on the hobby or activity. Address the following in your report:

1. Identify the chemistry knowledge and skills needed to participate in this hobby or activity. This must be included in the report.
2. Describe the role, if any, that technology plays in this hobby or activity. This must be included in the report.
3. Report the approximate number of people that participate in this hobby or activity. Include this information as available.
4. If references are used, they must be properly cited.

_____

_____

_____

58 **Scientists, Nature of Science, and History of Science Report**

Chemists have made significant contributions to human history. Select an important chemistry concept, idea, or theory and identify scientists that contributed to the development of the concept, idea, or theory. Write a 2 to 3 page, double-spaced report on the scientists and topic selected. Address the following in your report:

1. Clearly identify the chemistry concept, idea, or theory selected. This must be included in the report.
2. Describe the role that scientists played in the development of this concept, idea, or theory. This must be included in the report.
3. Describe how the historical development of the concept, idea, or theory illustrates the nature of science. This must be included in the report.
4. If references are used, they must be properly cited.

59   **Solubility Questions and Problems**

Note: To receive full credit on problems, you must show your work.

1. At a specific water temperature, 35.7 g of sodium chloride will dissolve in 100 cm^3 or 100 mL of water. Recall for our purposes, under certain conditions 1 g of water = 1 mL of water = 1 cm^3 of water. In a 244.3 g saturated solution of saltwater, what is the mass of the solute and the solvent?

2. A student investigation of the solubility of potassium nitrate at different water temperatures produced the following results:

Temperature (°C)	Maximum Amount Dissolved (g) in 100 cm^3 of Water
100.0	242.500
80.0	168.000
70.0	135.000
30.0	43.000
20.0	30.000
0.0	14.000

   a. Construct a solubility graph using graph paper provided by your instructor.
   b. How does temperature affect the solubility of potassium nitrate?

3. Use the graph you constructed in question 2 to answer the following:

A volume of 100 cm^3 of water at 100 °C is saturated with potassium nitrate is cooled. Fill in the table with the mass in grams of the sodium nitrate remaining in solution and the mass of the solid that precipitates from the solution

Temperature (°C)	Mass of solid remaining in solution	Mass of solid that will precipitate
100		
90		
50		
15		

4. From a variety of sources, a student was made aware of the following information:
   a. At close to room temperature, about 1.6 mg of calcium carbonate will dissolve in 100 cm^3 of water.
   b. A 500 mg tablet of calcium carbonate provides 200 mg of elemental calcium.
   c. The minimum daily requirement for calcium is 1000 mg.
   d. It has been suggested that to prevent osteoporosis, a woman should have between 1200 mg and 1500 mg of calcium per day.

For the purposes of this problem assume a woman has to dissolve calcium carbonate in water and drink the solution to obtain the daily-required amount of calcium (rather than swallowing tablets). How much water with calcium carbonate would she have to drink to obtain 1200 mg or 1.200 g of calcium in one day? Would you suggest the woman drink this amount of water during breakfast? Why or why not?

60   **Thermal Energy Questions and Problems**

Note: To receive full credit on problems, you must show your work.

1. When objects at different temperatures are brought together, what happens to their respective temperatures?

2. Graphs of data on the thermal expansion of liquids show that each liquid expands at a different rate as the temperature increases. Make a hypothetical sketch of these phenomena. Graphs of data on the thermal expansion of solids show that each solid expands at a different rate as the temperature increases. Make a hypothetical sketch of these phenomena. Graphs of data on the thermal expansion of gases show that each gas expands at about the same rate as the temperature increases. For gases, a best-fit line can be drawn. Make a hypothetical sketch of this phenomenon. You should have three sketches of graphs – one for liquids, one for solids, and one for gases.

3. Based on information provided in question two and accurate sketches of the graphs, can thermal expansion be used to identify and distinguish liquids? Solids? Gases?

4. Explain why most substances expand when heated using the kinetic theory of matter.

5. What must happen to the density of a solid when its temperature is increased? Explain.

6. Ten grams of steam at 105 °C is cooled to form ice at a temperature of –5 °C. Sketch a cooling curve on a properly labeled graph. Calculate the heat energy released using specific heats provided elsewhere in the text.

7. Explain how an alcohol thermometer works.

(Note: See other Activity-Based Physical Science units for additional reading on selected physical science concepts.)

Introduction

Chemistry is the study of matter. This includes the study of the composition of substances, their properties, and changes they undergo when they react with other substances.

One of the early motivations for studying matter certainly was human curiosity. Over time, another major reason has been to improve human existence by modifying natural materials. Technology can be used to apply knowledge gained through science for practical purposes. The development of new chemical products involves risks and benefits. Sometimes, there are unanticipated, unintended consequences when new chemical products are developed and used.

Matter

In earlier reading, physical science is generally described as the study of matter and energy. Matter has been described as having mass and taking up space. Mass is the amount of matter. Energy is often described as the ability to do work. Energy can cause changes in matter, either physical changes or chemical changes, to occur.

All matter can be classified as varying in composition (heterogeneous) or having the same composition throughout (homogeneous). Most of the matter found around you is made up of two or more different materials to form a **mixture**. Parts of a mixture retain their own properties. Can you think of a mixture? You may have suggested a fabric made of a blend of cotton and polyester. A mixture that is the same throughout is considered to be homogeneous, while a mixture that is not distributed evenly is considered to be heterogeneous. Air and tossed salads are examples of heterogeneous mixtures. Heterogeneous mixtures can be separated by physical means, i. e., by methods that would not change the properties of the constituents. Soda or pop is an example of a homogeneous mixture. It is made up of two or more materials that are evenly distributed. You would not want one swallow to have more sugar or more carbonation than another. Solutions (see section that follows) are examples of homogeneous mixtures. Some matter has the same composition throughout and is only made of one material. While homogeneous, this type of matter is not a mixture. **Elements**, composed of one kind of atom, and **compounds**, composed of two or more atoms of different elements, are homogeneous.

Atoms, Ions, Elements, and Compounds

All forms of matter consist of atoms. Some ancient scientific thinkers suspected that everything consisted of tiny particles. Democritus (about 470-400 BC) is given credit for suggesting that matter is made of tiny particles so small that they are invisible to human eyes. In addition, he believed the particles were in motion, lasted forever, and were very hard. Some parts of modern atomic theory match Democritus' ideas and some do not.

Atoms are the main building blocks of matter. There are about 112 different kinds of atoms. Different kinds of atoms make up different chemical **elements**. Some names of chemical elements are familiar, like gold, silver, or aluminum, while others are less so, like tantalum, erbium, and ytterbium.

Atoms are made up of even smaller parts generally called subatomic particles. The three main subatomic particles are protons, electrons, and neutrons. All electrons are the same. All protons are the same. All neutrons are the same. Various elements differ in the numbers of subatomic particles they have in each atom.

The British scientist, Ernest Rutherford, by 1911 had carried out experiments that showed atoms were made up of even smaller particles. He suggested a small, relatively heavy nucleus that had even smaller, lighter electrons orbiting around the nucleus. Niels Bohr, in 1913, improved upon Rutherford's ideas by suggesting that electrons had to stay at certain distances from the nucleus in layers called shells. An atom has a **nucleus**, which is the central part of an atom. The nucleus includes **protons**, which have a positive charge, and may contain **neutrons**, which have no charge. The number of protons defines the element. Neutrons and protons are about the same size; the mass of a proton is about 1.763×10^{-24} grams and the mass of a neutron is about 1.765×10^{-24} grams. **Electrons** are negatively charged particles that move around the nucleus in layers known as electron shells. The outermost electrons greatly affect the chemical reactivity of an element. Electrons located in outer shells have more energy than inner shell electrons. Overall, atoms have no electrical charge because they have the same number of electrons (negative charges) as protons (positive charges). Electrons have a much smaller mass than protons or neutrons. Electrons have a mass of about 9.11×10^{-28} grams. Isotopes are atoms of the same element but with different numbers of neutrons. Carbon-12 (6 protons, 6 electrons, 6 neutrons) and carbon-14 (6 protons, 6 electrons, 8 neutrons) are isotopes with carbon-14 being radioactive.

An atom that has gained or lost one or more electrons is called an **ion**. If electrons are lost from an atom, the ion will have a positive overall charge and be called a **cation**. If electrons are gained, the ion will have a negative overall charge and is called an **anion**. A **polyatomic ion** is made of two or more elements acting as one ion.

Compounds are substances containing two or more different elements chemically combined. They are made of specific ratios of different chemical elements. For example, water and hydrogen peroxide are both made up of hydrogen and oxygen. Water has two hydrogen atoms per one oxygen atom (H_2O) while hydrogen peroxide has two hydrogen atoms per two oxygen atoms (H_2O_2).

Elements and the Periodic Table

John Dalton (1766-1844) was a British science teacher and thought that every chemical element consisted of atoms that were identical to each other but different from the atoms of other elements. He gave names and symbols to about 30 elements. Dalton was not always correct with regard to his ideas about atoms and elements. Some materials he thought were elements are now known as compounds. Also, he thought atoms were solid spheres that could never be destroyed.

Dmitri Mendeleev (another spelling Mendelayef) proposed the periodic table of elements in 1868. He wrote down properties of the elements and then tried arranging them in different patterns. The best arrangement was one that showed how the elements in vertical columns have similar properties.

Of the 112 or so elements, around 90 occur naturally. The others have been made in laboratories. Elements in the periodic table are grouped according to similarities. Elements are arranged by the increasing number of protons in their nuclei.

Bonding

Atoms can join together, or **bond**, in several ways. Ionic bonds can form as the result of atoms losing or gaining electrons. Loss or gain of electrons may happen, for example, when an atom of a substance dissolves in a liquid. If an atom gains an electron it becomes a negative ion and can form an **ionic bond**. **Covalent bonds** occur when atoms share one or more electrons. Different levels or shells of electrons can hold up to a certain number of electrons. The innermost shell can hold up to 2 electrons and, in order, the next shells can hold 8, 18, 32, 32, 10, and 2 electrons. Electronic configurations are more complex than what is being described here. More details can be found in a chemistry course or chemistry text. Elements whose outermost electron shells are filled with electrons or hold eight electrons are stable and do not react as readily as elements with other electron configurations. If the outermost shell is not quite full of electrons, an atom can sometimes "use" electrons from another atom and hold onto them for part of the time. Similarly, if the outer shell has just one electron, it can donate that electron to another atom for part of the time. Atoms that share one or more electrons form a covalent bond.

Solutions

A **solution** is a homogeneous mixture. Two or more substances are uniformly mixed in a solution. The **solute** is the substance being dissolved and, usually, is the lesser quantity. The **solvent** is the substance doing the dissolving and is usually the greater quantity. In saltwater, water is the solvent and salt is the solute. Knowledge of solvents is useful when trying to remove stains from clothing or carpet. **Solubility** is the amount of solute that can be dissolved in a given amount of solvent at a specific temperature. Solubility is measured as the number of **grams** that will dissolve in **100 cm^3** of water (**g/100 cm^3** of water). A material is **insoluble** when it does not dissolve in a solvent.

Most persons think of solutions as **liquid solutions**. The solvent is a liquid and the solute can be a solid, liquid, or gas. Examples of liquid solutions include: soft drinks (sugar and carbon dioxide in water) and antifreeze (ethylene glycol in water). Water is the most common solvent. Dissolving solids, liquids, or gases in solids forms a **solid solution**. White gold is a solid solution of gold and palladium. Other examples of solid solutions are: sterling silver (copper in silver), steel (carbon in iron), and dental fillings (mercury in silver). Earth's atmosphere is a **gaseous solution** with oxygen and other gases dissolved in nitrogen. Humidity is water dissolved in air.

In the case of a solid dissolved in a liquid like salt in water, visualize the **solution process** as beginning on the outside surface of the salt crystals. The outer layer of salt particles is pulled away by the water. The next layer of salt now in contact with the water is stripped away and mixes with the water. This continues until all the salt is dissolved or until the water cannot dissolve any more salt. There are three ways to **increase the rate** at which a **solid dissolves** in a **liquid**: 1) stir or shake the solution, 2) heat the solution, and 3) increase the surface area of the solid solute. Cooling a solution with a solid solute may cause the solute to settle out of the solution. A **precipitate** is a solid that settles out of a solution. In the case of a gas dissolved in a liquid, increasing the temperature of the liquid decreases its ability to hold the gas in solution. An example of this is the solubility of oxygen in water; as the temperature of the water increases the amount of oxygen it can hold decreases. Ponds or small lakes sometime experience fish kills during the summer because of the lack of oxygen in the water.

A change in temperature may occur during the solution process. As particles of the solute pull away from each other and attach to the solvent, energy may be released (solvent cools off) or absorbed (solvent warms up). Instant ice packs are an example of where the temperature of the solvent decreases.

Concentration

Concentration is a measure of the amount of solute dissolved in a solution. Lots of solute dissolved is referred to as "**concentrated**" and little solute dissolved is referred to as "**dilute**." **Molarity** is the number of moles of solute in one liter or solution. What is a mole? Just as one dozen refers to 12 of something, a mole represents a number. A mole represents a very big number, an incredibly big number! A mole of atoms or molecules is about 6.022×10^{23} atoms or molecules. Written in regular notation this is:

$$602,200,000,000,000,000,000,000$$

When the atomic mass of any element is expressed in grams, the number of atoms in this amount is about 6.022×10^{23} atoms. This number is sometimes referred to as Avogadro's number. The value of Avogadro's number has changed slightly over time.

Chromatography

Most inks and dyes are a mixture of different colored substances (pigments) in a solvent. The process of chromatography can be used to separate the different pigments. **Chromatography** separates mixtures by percolation through an adsorbing medium. Yes, adsorbing is different from absorbing. Check into the difference if you are curious. The result can be stratified layers of different colors.

Chemical Reactions

A **chemical equation** is used to show changes that occur during a chemical change or reaction. Chemical equations use **symbols** for elements and **formulas** with number subscripts to show the ratio of elements in a compound. Substances to the left of the arrow are **reactants**. The arrow stands for produces or yields. Substances to the right of the arrow are **products**. Other symbols written as subscripts are used to indicate the states of the reactants and products. Commonly used symbol subscripts are: c for crystal and s for solid, l for liquid, g for gas, and aq for aqueous or dissolved in water.

Equations are balanced to show that matter is not lost nor created during a chemical reaction. Elements are arranged to produce new substances. Just because substances are brought together does not mean that a chemical reaction will occur. To balance a chemical equation, the formulas must be correctly written. In addition,

coefficients, or numbers written in front of the symbols and formulas, are used to keep the numbers of atoms in balance.

When heat energy is released during a chemical reaction, the reaction is called **exothermic**. Most chemical reactions people tend to notice are exothermic. Burning (sometimes called combustion) is an exothermic reaction. When heat energy is added or used during a reaction, the reaction is **endothermic**. Photosynthesis is an example of an endothermic reaction.

Catalysts are substances that change the rate of reactions without being permanently changed themselves. By contrast, **inhibitors** slow down reaction rates. Food preservatives are inhibitors added to foods to slow down the decay of food by preventing or reducing the growth of bacteria.

Chemical reactions can be classified in a variety of ways. **Synthesis** reactions occur when two or more elements or compounds unite to form one compound. Examples of synthesis reactions follow:

$$C_{(s)} + O_{2(g)} \rightarrow CO_{2(g)}$$

$$4 \, Fe_{(s)} + 3 \, O_{2(g)} \rightarrow 2 \, Fe_{(2)}O_{3©}$$

Decomposition reactions represent the breakdown of one substance into two or more other substances. Examples of decomposition reactions follow:

$$2 \, H_2O_{2(l)} \rightarrow 2 \, H_2O_{(l)} + O_{2(g)}$$

$$2 \, H_2O_{(l)} \rightarrow 2 \, H_{2(g)} + O_{2(g)}$$

A **single displacement** or **replacement** reaction is where one element displaces another in a compound. An example of a single displacement reaction is:

$$Cu_{(s)} + 2 \, AgNO_{3(aq)} \rightarrow 2 \, Ag_{(s)} + Cu(NO_3)_{2(aq)}$$

In a **double displacement** or **replacement** reaction the positive part of one compound unites with the negative part of another compound.

$$AgNO_{3(aq)} + NaCl_{(aq)} \rightarrow AgCl_{(s)} + NaNO_{3(aq)}$$

Acids, Bases, and Salts

Acids react with metals and contain hydrogen. Acids produce hydronium ion (symbol: H_3O^+) in water solution. Acids have a sour taste for those acids that can be tasted. Caution – do not taste acids unless directed to do so by your instructor. Proper safety precautions must be followed when handling acids.

Bases dissolve fats and oils, contain oxygen and hydrogen, and feel slippery. Bases produce hydroxide ion (symbol: OH^-) in water solution. Bases usually taste bitter for those bases that can be tasted. Caution – do not taste bases unless directed to do so by your instructor. Proper safety precautions must be followed when handling bases.

Acids or bases are substances whose relative strengths can be compared by using **neutralization reactions**. In the case of comparing bases, a standard acid is used to neutralize equal volumes of bases. **The base that requires the most acid to be**

neutralized is considered the strongest by comparison. As acid is carefully added, the base combines with the acid to produce a salt and water. The solution changes from being basic to being neutral after the addition of a sufficient amount of acid. After being neutralized, one more drop of acid added to a neutral solution would turn the solution acidic. In the case of comparing acids, a standard base is used to neutralize equal volumes of acids. **The acid that requires the most base to be neutralized is considered the strongest by comparison**. As base is carefully added, the acid combines with the base to produce a salt and water. The solution changes from being acidic to being neutral after the addition of a sufficient amount of base. After being neutralized, one more drop of base added to a neutral solution would turn the solution basic and, in the presence of the indicator phenolphthalein, give the solution a "pinkish" color.

In a neutralization reaction, the strength of an unknown acid or base can be determined using:

$$M_1V_1 = M_2V_2$$

M_1 and V_1 represent the volume of one solution and the molarity of that solution. M_2 and V_2 represent the volume of the other solution and the molarity of that solution. **Molarity** is a concentration unit with the symbol, **M**. Molarity is the number of moles of solute in one liter of solution.

An acid combined with a base yields a "**salt**" plus water. A chemical definition of salt goes beyond that of table salt, although table salt is chemically a salt. Salts have a brackish taste. A salt is an aggregate of positive and negative ions other than hydrogen and hydroxide ions.

The **pH scale** is a way to measure whether a solution is acidic, basic, or neutral. Originally, "pH" was an abbreviation for "potential of hydrogen ion." The pH scale has the advantage of expressing small concentrations of hydronium ion, H_3O^+ (think of H_2O combined with H^+), more conveniently than a decimal scale and does not make use of negative powers of 10. A neutral solution has a pH of exactly 7. Acidic solutions have a pH of less than 7 with 0 being most acidic and basic solutions have pH values more than 7 with 14 being the most basic. A one unit change on the pH scale equals a 10 times change in the strength of the acid or base. Thus a pH of 4 is 10 times more acid than a pH of 5 and 100 times more acid than a pH of 6. A pH of 14 is 10 times more basic than a pH of 13 and 100 times more basic than a pH of 12. Near 0 pH solutions will burn skin. Basic solutions with a pH near 14 will be harder on skin than near 0 pH acid solutions. For reference, here are the pHs of some "everyday" acids: battery acid has a pH near 0, stomach contents or gastric juices have a pH around 1 to 1.6, soft drinks have a pH of around 3, tomatoes 4, rainwater around 5.6 or 5.7, and milk 6. Some pHs of "common" bases include: blood about 7.35, baking soda 9, household ammonia about 11, and oven cleaner more than 13.

Acid Rain

Rainwater is normally slightly acidic, pH about 5.6 or 5.7 (depending on the source) as a result of passing through atmospheric carbon dioxide. Precipitation that falls through air polluted with sulfur oxides and nitrogen oxides, produces sulfuric and nitric acids. The pH of these **acid rains** (or other forms of precipitation) may be acidic enough to corrode materials and be harmful to various life forms. Acid rain pH can be as low as pH = 3.0. Burning fossil fuels contributes to the production of acid rain as do natural processes like forest fires and volcanic eruptions.

Indicators

Acid-base **indicators** are substances that turn different colors in acidic or basic solutions. Indicators have a dye that is sensitive to change in hydrogen ion concentration. **Litmus paper** and **pHydrion paper** are two examples of indicators. Red litmus paper will turn blue in a base solution. Blue litmus paper will turn red in an acid solution. If a solution does not change the color of red and blue litmus paper, then the solution is neutral. There are a number of other indicators such as phenolphthalein, bromothymol blue (BTB), and red cabbage extract. Red cabbage extract is pink to red in acids and blue to green in bases. BTB is "yellowish" (green to yellow) in acids and blue in bases. Phenolphthalein is an indicator that is colorless in an acid and turns pink in a base. Methyl red is red in an acid and yellow in a base.

Water

Water has a relatively simple chemical formula. If we could see atoms and molecules, water would be spherical in shape with two hydrogen atoms appearing as tiny bulges on a much larger oxygen atom. Water is a stable compound that does not break down easily. The table that follows illustrates the distribution of water on Earth.

Figure 4-5   Distribution of Earth's Water Supply

Location	Approximate Percentage
Saltwater or saline water in oceans	97.5
Freshwater in glaciers and ice caps	1.8
Freshwater as ground water	.63
Freshwater in lakes and rivers	.01
Freshwater in the atmosphere	.001

Water quality is a crucial factor in how water can be used. The taste of water depends on the type and amount of dissolved substances. Dissolved substances may originate from natural sources or from the activities of humans. Dissolved substances can be beneficial, as in the case of fluoride that helps prevent tooth decay. Bacteria from sewage and dissolved lead and arsenic can make water unsafe to drink.

Water Pollution

The dissolving power of water gives credence to its title as the "universal solvent." However, water does not dissolve all substances. Sometimes materials dissolved by water are pollutants.

Biodegradable wastes are wastes that can be broken down into harmless chemicals by organisms. Although the chemicals produced by the bacteria from breaking down biodegradable wastes are harmless, the environment can still be adversely affected. In the process of breaking down garbage, aerobic bacteria multiply and consume lots of oxygen. If oxygen levels drop too much, only rough fish can live in the lower oxygen levels. Even lower oxygen levels may cause all the fish and even the aerobic bacteria to die. Anaerobic bacteria can survive in low levels of oxygen but they are inefficient disposers of garbage that produce smelly gases. Biodegradable wastes can support beneficial bacteria and/or disease-causing bacteria. Sewage treatment plants can use chlorine to kill harmful bacteria and ozone can be used to kill viruses that cause certain diseases. Ozone is more expensive than chlorine. Water contains **dissolved oxygen** when oxygen dissolves into the water from natural or human activities. Examples include when oxygen from air dissolves into water as a product of photosynthesis, aeration of water as it moves over rapids and falls, or by human-made aerators. The amount of dissolved oxygen is lower in warm water, water with lots of bacteria, or water with lots of oxygen consuming aquatic animals. Run-off containing fertilizers can lower dissolved oxygen. Plants in fertilized water increase in size and number. Cloudy conditions cause respiring plants to use dissolved oxygen. Also, when plants die, bacteria will use dissolved oxygen as mentioned earlier.

An inverse relationship exists between the amount of dissolved oxygen in water and the water temperature. As water temperature increases, the amount of dissolved oxygen decreases. **Heat** or **thermal pollution** can occur in shallow bodies of water during the summer or when industries dump water used to remove heat energy into lakes or rivers. Thermal pollution may cause fish populations to switch to species more tolerant of warm water. A fish kill can result if water is heated to the extent that it does not hold enough oxygen.

Nitrogen makes up about 78% of the air in the atmosphere. Inorganic nitrogen can exist as a gas, as **nitrite** (NO_2^{1-}), **nitrate** (NO_3^{1-}), or **ammonia** (NH_4^{1+}). Nitrites are converted to nitrates by bacteria. Nitrites react with hemoglobin in the blood to produce methemoglobin. Methemoglobin destroys the ability of blood cells to transport oxygen. In babies under three months of age, this may result in "blue baby" disease. Organic nitrogen is found in proteins and other compounds. When fertilizers and other products that provide plant nutrients get washed into streams and lakes, water plants grow at faster rates. As mentioned earlier, this ultimately can reduce oxygen levels. Nitrate is a major ingredient in fertilizers used for agricultural purposes and for lawns. Other sources of nitrates include: animal wastes, car exhaust, and leaking septic tanks.

While other elements may contribute to hard water, **hardness** is defined as the amount of calcium and magnesium dissolved in water. Usually, the other elements are not found in large amounts naturally. Carbon dioxide combines with water to form a weak solution of carbonic acid. Carbonic acid eats away at limestone and produces calcium carbonate – a white compound that contributes to hard water. Calcium carbonate sometimes builds up in teakettles as a white, scaly compound. The total hardness test measures the total amount of calcium and magnesium in the water. It is measured in SI units of mg/L or USCS/English units of grains per gallon (gpg). To convert mg/L to gpg, divide mg/L by 17.1. To convert gpg to mg/L, multiply gpg by 17.1.

Silt can be a major water pollution problem. Silt is made up of very small particles of soil or rock and gets washed into lakes and streams after heavy rains or snow melt. Cloudy or turbid water decreases the penetration of sunlight into water and affects photosynthesis. Sediment, silt that settles to the bottom, in large amounts can change the nature of the bottom and interfere with the reproductive cycle of fish.

Some poisonous wastes ingested in the body do not break down. The process where additional poison builds upon the poisons already accumulated is called **bio-accumulation**. Infamous examples of substances that bio-accumulate include mercury, DDT, and chlordane. By chewing on objects painted with lead based paints, babies and small children have been affected by lead poisoning.

Chemistry and Society

The important role that chemistry plays in society and personally cannot be disputed. A few examples of the role that chemistry plays will be discussed here.

Organic chemistry is the chemistry of carbon compounds. Initially, the term "organic" meant derived from living things. At one time it was thought that organic compounds were found only in organisms. As chemists studied the chemical makeup of living things, they found that carbon was found in most of the organic compounds. Organic chemistry became the chemistry of carbon compounds. However, there are compounds that contain carbon that are not considered organic. Carbon dioxide (CO_2) and limestone ($Ca\ CO_3$) are examples. Plastic and other synthetic carbon-containing compounds are a part of organic chemistry. Over 90% of all known compounds are organic. Other than water, most of the chemicals that make up the human body are organic. Most fuels people use, the food people consume, the many plastics used daily, and many types of clothing are organic compounds.

Food serves as a source of energy for the body and provides the raw material for growth and maintenance of the body. Food helps control and facilitate biological and chemical processes that occur in the body. Chemicals taken in as food or drink include: **water** that serves as a solvent; **proteins** help build and maintain the body; **carbohydrates** made up of carbon, hydrogen, and oxygen serve as an energy source; **fats** do not easily dissolve in water so serve as a way of storing energy; and, **vitamins** and **minerals** are necessary for good health.

Recycling involves the recovery and reuse of Earth's resources. Although Earth receives energy from the Sun and releases energy to space, the Earth system is nearly a closed system when it comes to matter. Relatively little matter is received from outside the Earth system and little matter escapes from the Earth system. Energy is used to power a number of natural cycles. Just as water is recycled in the water cycle, nature recycles other atoms and molecules of matter on Earth. Carbon is an example of an element that is used and reused in different forms.

Recycling of materials by people first involves collection of the materials. Then the "trash" is separated into what can be recycled ("recyclables") and what cannot be recycled. Recyclables are cleaned up and then processed so they can be reused. Aluminum cans contain about 97% aluminum with small amounts of magnesium, manganese, iron, silicon, and copper. Recycling aluminum cans uses much less energy that it would take to manufacture the cans from aluminum ore called bauxite. Besides aluminum, many communities have programs to recycle plastics, newspaper, and paper.

Heat and Temperature

Hotness or coldness is indicated by **temperature**. Temperature is a measure of the average speed or kinetic energy of molecules or atoms of a substance as measured using a thermometer. Temperature can be measured in **Celsius degrees, Fahrenheit**

degrees, or **Kelvin** units. **Heat** is the energy of a system caused by the continuous motion of its particles and takes into account temperature and mass.

Absolute zero is set as zero on the Kelvin scale and is the coldest theoretically possible temperature. At this temperature, it is theorized that particles of matter would have the least motion (at one time the speculation was no motion).

A **calorie** is a unit of energy that represents the amount of heat needed to raise the temperature of one gram of water by one degree Celsius. A **Calorie** is equal to 1000 calories and is a kilocalorie or 1 dietary calorie.

Kinetic Theory of Matter

Particles of matter are in constant motion. This is known as the **Kinetic Theory of Matter**. During evaporation, a liquid absorbs heat energy from its surroundings. Heat raises the average kinetic energy of the particles. Some of the liquid particles may gain enough kinetic energy to escape the surface of the liquid to become gaseous particles. Energy is required to change particles from a liquid to a gaseous phase or state. Evaporation can occur without the water changing temperature as when "room temperature" evaporates

Heat energy drives the evaporation process. The average kinetic energy is higher for a gas than for a liquid. This is obviously true when the gaseous phase is at a higher temperature than the liquid phase, but is also true when the phases are at the same temperature. For example, with each at room temperature, gaseous H_2O has more kinetic energy than does liquid H_2O. Temperature is a measure of the average kinetic energy of the molecules (or other particles) of a substance.

A number of household appliances designed for cooling use the process of evaporation. Historically, freon has been a commonly used substance in these appliances. Concerns about the effects of freon use on the ozone have caused manufacturers to use alternatives to freon.

Phase Changes

Matter can change phase or state from solid to liquid to gas when energy, e. g., heat, is added or the reverse when energy is removed. However, you have observed that water (liquid) evaporates and changes to a gas (water vapor) at "room temperatures" without dramatic changes in temperature.

Under defined conditions **boiling point** is the temperature at which a substance changes rapidly from a liquid to a gaseous state. **Boiling point elevation** refers to the phenomena by which the boiling point temperature will be raised when an impurity (another substance) is added. An example of this is when water, the impurity, is added to coolant or antifreeze in automobile radiators. Another example is adding salt to water when cooking. The combination of salt and water will boil at a higher temperature than water alone. Under defined conditions **freezing point** is the temperature at which a substance changes from a liquid to a solid. **Freezing point depression** refers to the phenomena by which the freezing point temperature will be lowered when an impurity is added to a substance. An example of this is the addition of salt or other commercial "deicer" to the ice formed on streets and sidewalks. The temperature is not raised to melt the ice, rather the freezing temperature is lowered so H_2O will exist as a liquid at a lower temperature than the usual freezing point temperature. **Melting point** is the temperature at which a solid changes to a liquid. Under the same conditions, melting point temperature and freezing point temperature are the same for pure substances.

The **heat of fusion** of a substance is the amount of heat required to change one gram of a solid to a liquid, i. e, to melt the solid. The **heat of vaporization** is the amount of heat needed to vaporize one gram of a liquid, i. e., to change the liquid to a gas. **Specific heat** or **heat capacity** is the amount of heat needed to raise the temperature of one gram of a substance by one degree Celsius.

Scientists

In 1756, a Scot named Joseph Black (1728-1799) discovered that when heated magnesium carbonate lost weight. He suggested that the substance gave off gas during the heating process and named the gas "fixed air." His fixed air is now known as carbon dioxide.

Born in England, Joseph Priestley (1733-1804) immigrated to the U. S. in 1791. While in England, Priestley made an important discovery. He noticed that the gas given off when mercuric oxide was heated caused a lighted candle to burn brighter. At the time of his observation many scientists believed that when something burned it lost a substance called phlogiston. Priestley called the gas "dephlogisticated air." The gas identified is now known as oxygen.

In 1661 Robert Boyle (1627-1691), also from England, proposed that elements are substances that cannot be broken down into simpler substances. This helped to clarify the notion of an element.

Antoine Lavoisier (1743-1794) worked as a tax collector in France to support his scientific research. He was executed near the end of the French Revolution. Lavoisier conducted a number of experiments investigating combustion or burning. His results showed that substances often became heavier after burning rather than losing weight. He thought the substances must have absorbed something from the air and showed this material to be the gas Priestley named dephlogisticated air. He renamed it oxygen. Lavoisier provided evidence to disprove the phlogiston theory which had been held for a hundred years. He is given credit for being the first person to demonstrate burning involves the addition of oxygen. Lavoisier published *Methods of Chemical Nomenclature* in 1789. In it he developed the idea of elements as chemical building blocks that cannot be broken down into simpler substances. The book introduced a system of naming substances based on their chemical content and listed 33 elements showing how they combined to form compounds. Lavoisier demonstrated that elements can combine chemically to form compounds and that compounds can be broken down into elements.

John Dalton (1766-1844), also English, published *A New System of Chemical Philosophy* in 1808. Three important ideas were discussed in this book. One was that all chemical elements are made up of atoms that do not break up during chemical reactions. A second was that all chemical reactions are the result of atoms combining or separating. A final major idea was that different atoms weigh different amounts.

Dalton is given credit as the first person to suggest a comprehensive theory of how atoms function as the building blocks of matter. Dalton built upon the ideas of others and contributed ideas of his own.

Dalton thought elements to be composed of tiny, indivisible particles or atoms. The ancient Greeks recognized two models of the nature of matter. The more popular model considered matter to be continuous and subject to being divided into smaller and smaller pieces. Aristotle agreed with the majority view that matter consisted of four elements, earth, air, fire, and water which possessed qualities of hot, cold, dry, and moist. It was thought that substances could be changed into other substances by modifying their qualities. These mistaken ideas, held for centuries, led many to attempt to change common metals to valuable ones like gold. Democritus and others held the less popular view that matter consisted of a number of small particles that combine in a variety of ways to produce a number of different substances. Democritus is given credit for the term "atom" which means indivisible.

Dalton thought atoms to be indestructible based on the experimental work of Lavoisier. Lavoisier developed a sensitive balance and used it to determine mass changes of chemical reactions in sealed containers. Lavoisier was not able to detect a change in the total mass during a chemical reaction. This is now known as the **Law of Conservation of Mass**.

Dalton postulated that atoms of a given element are identical in character and atoms of different elements are different in character. Today, we recognize that atoms of the same element can have different numbers of neutrons (isotopes) which affect mass but do not affect chemical behavior. The outermost electrons control chemical behavior. Atoms of different elements are different in size due to their different electron structures.

Another Dalton postulate was that atoms of different elements join together to make what are now called molecules. Dalton prepared a chart showing known elements and compounds. He presented the structures of some simple molecules although he erred with the formula of water thinking it to be HO instead of H_2O. Later, others conducted experiments and their results corrected this error.

Dalton proposed that different kinds of atoms in a compound are present in simple numerical ratios. In the late 1700s a French chemist, Joseph Proust, and others conducted experiments that contributed to the **Law of Definite Proportions**. The Law of Definite Proportions states that a compound always contains two or more elements combined in a definite proportion by mass. In 1809, another French chemist, Joseph Gay-Lussac, experimentally determined that two volumes of hydrogen gas combine with one volume of oxygen gas to produce two volumes of water vapor. Shortly after Gay-Lussac's discovery in 1811, the Italian physicist Amadeo Avogadro claimed that under identical conditions of temperature and pressure, equal volumes of any kind of gases contain the same number of molecules. This assertion is known as Avogadro's Law.

Also, Dalton discovered that atoms of two or more elements can combine in different ratios to produce more than one compound. This is called the **Law of Multiple Proportions.** This law supports the idea that compounds are made of more than one kind of atom (element) and the idea that these elements can combine in different ratios.

Dmitri Mendeleev (1834-1907) was born in Siberia and later became a professor of chemistry at the University of St. Petersburg. In 1869 he published his Periodic Table of the Elements. He grouped elements into families according to their atomic weights, smallest (hydrogen) on the left and largest (lead) on the right. His periodic table illustrated how the elements are related to one another. Mendeleev had gaps in his periodic table that he claimed were elements yet to be discovered which was correct. The modern periodic table arranges elements according to atomic number (the number of protons or electrons) which produces some slight changes compared to listing elements by increasing atomic weight.

Science Careers

Chemistry is the scientific study of matter or substances including their composition, structure, properties, and interactions. Chemists have made many important contributions to society. Chemistry careers provide challenging opportunities in education, industry, and government. As was stated in the More Physical Science Reading, chemists are employed in a variety of roles including research and development, teaching, materials and product testing, production, and sales. Chemists are employed by government agencies and in the private sector. Many health-related careers require chemistry background. Doctors of medicine and pharmacists are two health careers that require in-depth knowledge of chemistry. Chemists are employed in a wide many related professions such as material science, molecular biology, forensic science, and hazardous waste management.

Specialty areas of chemistry include: analytical, biochemistry, inorganic, organic, and physical. A college degree is required and many jobs require an advanced degree. For more information about chemists and chemistry, contact the American Chemical Society.

Chemistry Case Study – Studying Fire and More

Science as a human endeavor is illustrated as early chemists studied combustion and in the process recognized conservation of matter, the importance of making careful measurements, and communicating their results.

Prehistoric people made useful discoveries by observing properties of natural substances and changes those substances undergo. For a long time people believed that combinations of a few basic materials make up all the substances that exist in the world. Around 400 B.C., the Greek philosopher Empedocles is credited with the theory that stated the basic substances were earth, water, fire, and air. Aristotle claimed one of the four basic elements could be changed into any of the others by adding or removing heat and moisture. The search for the basic kinds of matter continues today.

French scientist, Antoine Lavoisier, did experimental and theoretical work in the time span between the American and French revolutions that led to the modern science of chemistry. His work in the late 1700s revolutionized chemistry. Lavoisier carefully measured all the substances involved in burning and provided evidence that supported the idea of conservation of mass or matter, i.e., the total amount of matter before a change is the same as the total amount of matter after a change. His quantitative methods and emphasis on physical laws and a theory of materials changed how scientists approached their work.

The chemical practice of alchemy was a major source of chemical knowledge. It started with Egyptian scholars during the first 300 years B.C., spread to other areas and continued into the 1600s. However, two of their main goals of alchemy were to change some metals to gold and to produce an "elixir of life," i. e., a substance that would allow people to live forever. Obviously, they failed in these two attempts. Lavoisier's scientific study of chemistry moved the discipline beyond alchemy and has provided greater understanding of chemical reactions and the production of new materials. Lavoisier's system for naming substances and describing their reactions allowed other scientists to communicate their findings with one another.

John Dalton made improvements to the ancient Greek ideas of element, atom, compound, and molecule. His atomic theory was based on the idea that each chemical element has its own kind of atoms. This benefited the modern development of chemistry by providing a physical explanation for chemical reactions that could be expressed quantitatively.

Chemistry Case Study – Splitting the Atom and More

The French physicist Antoine Henri Becquerel accidentally discovered that rays coming from a uranium ore affected a photographic plate in a manner similar to x-rays or light. His suggestion led Marie and Pierre Curie to work with the pitchblende and to eventually isolate an element they named radium. Also, they isolated polonium named for Marie Curie's country of birth. Radium and polonium were separated from tons of uranium ore, pitchblende. They identified these two elements in their search to account for the source of radioactivity beyond what could be accounted for by the uranium itself. Marie Curie shared a Nobel prize in physics with her husband and later won a Nobel prize in chemistry. She was the first scientist to win two Nobel prizes. The study of radium by the Curies and other scientists eventually led to the conclusion that one kind of atom can change into another kind.

Ernest Rutherford, from New Zealand, and colleagues developed the idea that radioactive uranium splits into a slightly lighter nucleus and a helium nucleus. Following up Rutherford's work, German and Austrian scientists showed that when struck by neutrons, uranium splits into two nearly equal parts and one or two extra neutrons. Lisa Meitner first suggested that if the mass of the fragments from such a change was less than the mass of the original uranium nucleus, then Einstein's special relativity theory predicted the release of a large amount of energy. Working with others in the U. S., the Italian scientist Enrico Fermi showed extra neutrons serve as atomic bullets to trigger the splitting of other nuclei and create a chain reaction in which large amounts of energy would be produced. Working in England in 1911, Rutherford developed a model of the atom, which called for a dense positive charge in the small, spherical nucleus with electrons travelling around the nucleus.

In 1938, German physicists Otto Hahn and Fritz Strassman discovered nuclear fission by splitting uranium atoms. The United States devoted massive amounts of resources to develop technology and in building nuclear fission bombs used in Japan near the end of World War II. The development of nuclear fusion weapons followed. Nuclear reactors were constructed to use nuclear energy to produce electricity. Besides weapons and in generating energy, radioactivity is used in medicine, industry, and to conduct scientific research.

absolute zero

acid

acid rain

alkalinity

atom

base

binary compund

bitter

boiling point

boiling point elevation

bromothymol blue (BTB)

bonding

calorie (food vs. science)

Celsius

change (chemical, nuclear, physical)

chemical equation (balance)

chemical reaction

chromatography

classifying

coefficient

colloid

color change

compound

concentrated

concentration

constellation

contract/contraction

cooling curve

degree

dilute

dissolved oxygen

dissolving

electrons (lose, gain)

element

endothermic

energy

environment

evaporation

exothermic

expansion

Fahrenheit

formula (chemical, writing)

freezing point

freezing point depression

gas

hardness (water)

heat

heating curve

heat of fusion

heat of vaporization

hydronium (hydrogen) ion

hydroxide ion

impurity

indicator

inorganic

interaction

ion

isotope

kinetic energy

kinetic theory of matter

Law of Conservation of Mass

Law of Definite Proportions

Law of Multiple Proportions

liquid

litmus paper

mass

matter

melting point

molecule

molarity

negative

neutral

neutralization (reactions; acid-base)

neutron

nitrate

organic

oxidation number

pH

phase change (or change of state)

phenolphthalein

pHydrion paper

polyatomic ion

positive

precipitate

proton

reaction

red cabbage

salt

saturated

slippery

solidify (solidification)

solute

solution process

sour

strong

supersaturated solution

system

thermal expansion

unsaturated solution

weak

scientific notation

solid

solubility

solution

solvent

specific heat

subscript

suspension

temperature

thermometer

water quality

writing chemical formulas

Name(s)   _____

_____

_____

_____

63   Scale Model of the Solar System

Related National Science Education Standards:	
Science Content:	systems, order, and organization
	evidence, models, and explanation
	constancy, change, and measurement
	evolution and equilibrium
Science as Inquiry:	abilities to do scientific inquiry
	understanding about scientific inquiry
Physical Science:	position and motion of objects
	motions and forces
	light, . . .
Earth and Space Science:	objects in the sky
	changes in Earth and sky
	structure of the Earth system
	energy in the Earth system
Science in Personal and Social Perspectives:	changes in environments
History and Nature of Science:	science as a human endeavor
	nature of science
	nature of scientific knowledge
	historical perspectives

Exploration/Awareness:

You probably have noted highway maps provide scales that allow users to convert distances on the map to "real world" distances. For example, if the map distance between two points is measured as two inches and the map scale is 1 inch = 25 miles, you know the "real world" distance is 50 miles.

In this activity you will be asked to construct a simple scale model of the solar system on adding machine paper. On one side of the paper you will construct a model that illustrates relative distances from the Sun to the planets. On the other side of the paper you will construct a model that illustrates relative diameters of the planets and the Sun. The scaled distance used to represent distances from Sun to planets will be different than the scale used to represent diameters of the planet and the Sun. At the end of the activity you should be able to provide a good reason why a different scale was used. Based on the data provided in the Planet Data table that follows and the skills developed in this activity, you should be able to use your knowledge and creativity to construct a three dimensional model of the solar system.

Table 5-1   Planet Data (data taken from NASA sources + others; discovery of additional moons requires updating)

Planet	Avg. Dist. from Sun (mill. km)	Avg. Dist. from Sun (AU)	Equat. Diam. (km)	Period of Rev.[1]	Period of Rot.[2]	No. of Moons	Appr. Dens. (g/cc)	Atmo-sphere	Other
Mercu-ry	58	0.4	4880	88 days	59 days	0	5.4	almost none; 0.0 bars	black sky; cra-ters
Venus	108	0.7	1.21×10^4	225 days	-243 days	0	5.2	CO_2; ~90 bars	"green-house effect"
Earth	150	1.0	1.28×10^4	365.25 days	23 h 56 min.	1	5.5	N_2, O_2; 1.0 bar	
Mars	228	1.5	6790	687 days	24 h 37 min.	2	3.9	CO_2, N_2, Ar; 0.007 bar	Olym-pus Mons
Jupiter	778	5.2	1.43×10^5	12 yrs.	9 h 50 min.	4 large; 39 or more	1.3	H, He, CH_4, NH_3	Great Red Spot, rings
Saturn	1427	9.5	1.20×10^5	29.5 yrs.	10 h 14 min.	18 named 30 or more	0.7	H, He, CH_4, NH_3	rings
Ura-nus	2870	19.2	5.20×10^4	84 yrs.	-17 h 54 min.	18 or more	1.2	H, He, CH_4	rings
Nep-tune	4497	30.1	4.95×10^4	165 yrs.	19 h 6 min.	8	1.7	H, He, CH_4	rings
Pluto	5900	39.4	2300	248 yrs.	6.4 days	1	uncer-tain, near 1.8	likely N, CH_4 CO,	

[1] period of revolution in Earth days or years; counterclockwise as viewed from above
[2] period of rotation in Earth days or hours and minutes; counterclockwise as viewed from above

Data presented in the table include: average distance from the Sun in millions of kilometers and in AU (astronomical units), diameter at the equator in kilometers, period of revolution (time to revolve around the Sun), period of rotation (time to spin once on its axis), number of moons, approximate density, atmospheric information, and "other" information.

On one side of a strip of adding machine paper a little longer than 200 cm, construct a scale model of the solar system that illustrates relative distances from the Sun to the planets.

Use a scale that has Pluto 200 cm from the Sun. To determine the scale to use, 200 cm is being used to represent 5,900,000,000 kilometers. What distance does 1 cm represent?

$$\frac{200 \text{ cm}}{5,900,000,000 \text{ km}} = \frac{1 \text{ cm}}{x \text{ km}}$$

The value of x is the "real world" distance that 1 cm represents.

What real world distance does 1 cm represent in this scale model of the solar system?

1. Using the relationship calculated in question one earlier, determine the scale model distances from the Sun to the planets:

    Mercury   _____ cm

    Venus   _____ cm

    Earth   _____ cm

    Mars   _____ cm

    Jupiter   _____ cm

    Saturn   _____ cm

    Uranus   _____ cm

    Neptune   _____ cm

    Pluto   _____ cm

2. On one side of the strip of adding machine paper, mark an "S" to represent the Sun. Note the relative distance to Pluto by making a mark 200 cm from the S and labeling it as Pluto. Similarly, mark and label the relative or scale distances to the other planets.

On the other side of the adding machine paper a little longer than 200 cm, construct a scale model that illustrates relative diameters of the planets and the Sun. Use a scale that has the diameter of the Sun equal to 200 cm. To determine the scale, 200 cm is being used to represent the diameter of the Sun. The diameter of the Sun is about 1.39 x 10^6 kilometers.

$$\frac{200 \text{ cm}}{1390000 \text{ km}} = \frac{1 \text{ cm}}{x \text{ km}}$$

The value of x is the "real world" distance that 1 cm represents or the diameters of the planets.

3. What real world distance does 1 cm represent?

4. Using the relationship calculated in question four, determine the scale model diameters of the Sun and the planets.

Sun	_____	cm
Mercury	_____	cm
Venus	_____	cm
Earth	_____	cm
Mars	_____	cm
Jupiter	_____	cm
Saturn	_____	cm
Uranus	_____	cm
Neptune	_____	cm
Pluto	_____	cm

5. On the opposite side of the strip of adding machine paper on which you marked the relative distances from the Sun to the planets, note the relative diameters of the Sun and the planets. Do this by making arcs for the Sun and the larger planets or circles for the smaller planets on the adding machine paper. Note: the relative diameters will NOT be drawn at the proper relative distance from the Sun because a different scale is being used for distances and diameters.

Concept Development:

6. What conclusions about the solar system can be drawn from observing the relative distances to the planets and the relative sizes of the Sun and the planets?

7. What does the "-" sign in front of the period of rotation for Venus and Uranus mean?

8. What is unique about the periods of revolution and rotation for Venus?

9. How do you account for the period of revolution and period of rotation data for Earth?

10. The inclination of the axes on which the planets rotate varies from planet to planet. Many know that the inclination of the Earth's axis or tilt is about 23.4° or 23.5°. Mercury, Venus, and Jupiter have small inclinations. Uranus has a tilt that is "around" 90°. The angle represents how far from the vertical the planets' axes are tilted from the vertical.
   a. Of the planets mentioned, which planet(s) spins on its axis "like a top?"

   b. Of the planets mentioned, which planet(s) spins on its axis "like a rolling ball?"

Application:

The consensus among scientists is the Earth is about 4.6 billion years old. Initially, using relative dating methods, scientists constructed the geologic time scale. Radiometric dating methods developed later have been used to apply absolute dates.

Time	Started	Ended
Tertiary period	~65 million years ago	present
Cretaceous period	~144 million years ago	~65 million years ago
Jurassic period	~202 million years ago	~144 million years ago
Triassic period	~245 million years ago	~202 million years ago
Permian period	~286 million years ago	~245 million years ago
Carboniferous period	~360 million years ago	~286 million years ago
Devonian period	~410 million years ago	~360 million years ago

Silurian period	~433 million years ago	~410 million years ago
Ordovician period	~505 million years ago	~433 million years ago
Cambrian period	~540 million years ago	~505 million years ago
Precambrian era	~4.6 billion years ago	~540 million years ago

11. If all of geologic time (4.6 billion years) was represented by 460 meters or 46000 centimeters, what amount of time would 1 cm represent?

12. Using the scale developed in #11, what distance or length would represent the start of the Triassic to the end of the Triassic period?

13. Using the scale developed in #11, what distance or length would represent the end of Precambrian time to the present?

14. If a basketball is used to represent the Sun, what object or size of object could be used to represent the Earth keeping the scale in perspective?

Background Information:
    For additional background, read the **Astronomy** and **Geology** sections of the Earth and Space Science Reading.

Name(s) _____

_____

_____

_____

64   **Moon Phases**

Related National Science Education Standards:	
Science Content:	systems, order, and organization
	evidence, models, and explanation
	constancy, change, and measurement
	evolution and equilibrium
Science as Inquiry:	abilities to do scientific inquiry
	understanding about scientific inquiry
Physical Science:	properties of objects and materials
	position and motion of objects
	motions and forces
	light, . . .
Earth and Space Science:	objects in the sky
	changes in Earth and sky
	structure of the Earth system
	Earth in the solar system
History and Nature of Science:	science as a human endeavor
	nature of science

Exploration/Awareness:

       You and/or members of your group have been making observations of the Moon; among other things noting its phase, relative location, and time of day. In this activity, you will use a simplified model to develop an explanation for why phases of the Moon occur.

       Make observations as if *you were standing on Earth* to complete the table. Use a protractor to make the Sun to Earth (as the vertex) to Moon angle measurements. The Sun to Earth side of the angle will always pass through position 1 on the model. All angle measurements should be made so they are equal to or less than 180°.

Table 5-2   Moon Phases

Moon position	Sketch of Moon as viewed from Earth	Phase	Sun-Earth-Moon Angle
1			

Moon position	Sketch of Moon as viewed from Earth	Phase	Sun-Earth-Moon Angle
2			
3			
4			
5			
6			
7			
8			
9			

Moon position	Sketch of Moon as viewed from Earth	Phase	Sun-Earth-Moon Angle
10			
11			
12			

Concept Development:

Use notes from the question and answer/lecture part of class to complete the "Phase" column of the table if you have not already done so.

13. In our simplified model, what is the angle requirement for a full Moon?

14. In our simplified model, what is the angle requirement for a crescent Moon?

15. In our simplified model, what is the angle requirement for a gibbous Moon?

16. In our simplified model, what is the angle requirement for a quarter Moon?

17. In our simplified model, what is the angle requirement for a new Moon?

18. Based on the model and observations from your Moon log, as viewed from above, does the Moon travel in a clockwise or counterclockwise path around Earth?

19. Elementary students often give the name "half Moon" to what we will call a "quarter Moon" because they see half of the "disk" or "circle" of the Moon illuminated. Give a good reason why this phase is called "quarter Moon."

20. In what numbered positions do you think an eclipse might occur? Why?

Application:
21. With help from your instructor as needed, explain why there are not as many eclipses as suggested by this model.

22. If directed by your instructor, go to an appropriate web site (such as NASA at http://www.nasa.gov/ or, if this URL is outdated, conduct a search to find the current URL) to read and study the explanations provided for phases of the Moon and eclipses.

Background Information:
     For additional background, read the **Astronomy** section of the Earth and Space Science Reading.

65   Time, Moon Phase, and Moon Location

Related National Science Education Standards:	
Science Content:	systems, order, and organization
	evidence, models, and explanation
	constancy, change, and measurement
	evolution and equilibrium
Science as Inquiry:	abilities to do scientific inquiry
	understanding about scientific inquiry
Physical Science:	position and motion of objects
	motions and forces
	light, . . .
Earth and Space Science:	objects in the sky
	changes in Earth and sky
	structure of the Earth system
	energy in the Earth system
Science in Personal and Social Perspectives:	changes in environments
History and Nature of Science:	science as a human endeavor
	nature of science
	nature of scientific knowledge
	historical perspectives

Exploration/Awareness:

In this activity you will use knowledge/understanding developed in the Moon Phase activity to determine one of the following given the other two: Moon location, Moon phase, or time (or the Sun's position). Although not always true but to keep a simple model will assume the Sun rises at 6:00 AM and sets at 6:00 PM.

As a review complete the following:

1. What conditions are necessary for a full Moon?   _____

2. What conditions are necessary for a new Moon?   _____

3. What conditions are necessary for a quarter Moon?   _____

4. What conditions are necessary for a crescent Moon?   _____

5. What conditions are necessary for a gibbous Moon?   _____

Concept Development:

Given two of Moon's location (marked by an "x"), Moon phase, or time (Sun's approximate position), determine the third.

6.

Moon Location	Moon Phase	Time
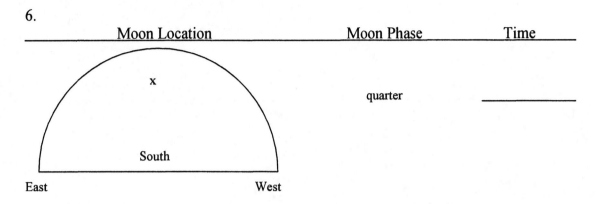	quarter	_____

7.

Moon Location	Moon Phase	Time
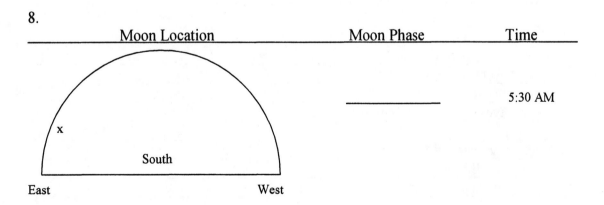	full	8:00 PM

8.

Moon Location	Moon Phase	Time
	_____	5:30 AM

9.

Moon Location	Moon Phase	Time

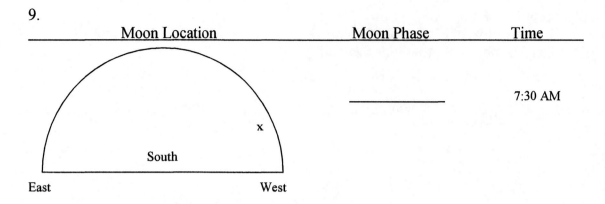

_____   7:30 AM

10.

Moon Location	Moon Phase	Time

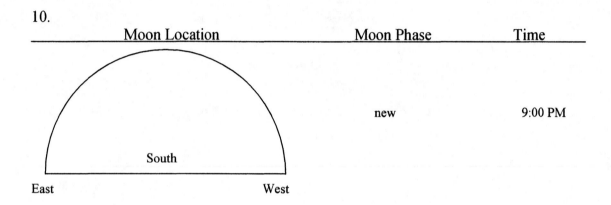

new   9:00 PM

Application:

11. Just as the Sun has sunrise and sunset times, the Moon has moonrise and moonset times. Again, based on knowledge and understanding developed in the Moon Phases activity and this activity, complete the table that follows.

Moon Phase	Moonrise Time
new	
"average" new crescent	
first quarter	
"average" new gibbous	
full	
"average" old gibbous	
third quarter	
"average" old crescent	

66   **Star Cycles Research Report**

Related National Science Education Standards:	
Science Content:	systems, order, and organization
	evidence, models, and explanation
	constancy, change, and measurement
	evolution and equilibrium
Science as Inquiry:	understanding about scientific inquiry
Physical Science:	light, . . .
Earth and Space Science:	objects in the sky
	changes in Earth and sky
History and Nature of	
Science:	science as a human endeavor
	nature of science
	nature of scientific knowledge
	historical perspectives

Stars go through a process where they form, exist, and then cease to exist. Use one or more properly cited references to prepare and word process a report that chronicles these stages of development. Your report must include appropriate mention of the Hertzsprung-Russell or H-R diagram. The report should be two pages, double spaced with 12-point font. Properly cited references should appear on the third page.

Name(s) _____

_____

_____

_____

67   **Planetarium and/or Observatory Visit**

Related National Science Education Standards:	
Science Content:	systems, order, and organization
	evidence, models, and explanation
	constancy, change, and measurement
	evolution and equilibrium
Science as Inquiry:	abilities to do scientific inquiry
	understanding about scientific inquiry
Physical Science:	position and motion of objects
	motions and forces
	light, . . .
Earth and Space Science:	objects in the sky
	changes in Earth and sky
	structure of the Earth system
	energy in the Earth system
Science in Personal and Social Perspectives:	changes in environments
History and Nature of Science:	science as a human endeavor
	nature of science
	nature of scientific knowledge
	historical perspectives

In this activity you will attend a local planetarium presentation and/or a local observatory. Write a summary of the important ideas presented.

68   **Simulating the Greenhouse Effect**

Related National Science Education Standards:	
Science Content:	systems, order, and organization
	evidence, models, and explanation
	constancy, change, and measurement
	evolution and equilibrium
Science as Inquiry:	abilities to do scientific inquiry
	understanding about scientific inquiry
Physical Science:	properties of objects and materials
	. . . , heat, . . .
	properties and changes of properties in matter
	transfer of energy
	conservation of energy . . .
	interactions of energy and matter
Life Science:	. . . and environments
Earth and Space Science:	properties of Earth materials
	changes in Earth and . . .
	structure of the Earth system
	energy in the Earth system
Science in Personal and Social Perspectives:	changes in environments
	. . . environments
	risks and benefits
	. . . human-induced hazards
History and Nature of Science:	nature of science

Exploration/Awareness:

Models are frequently used in earth and space science when the natural phenomena being studied is too large, too small, or too complex to easily be studied in the physical science classroom. Models will vary in the degree to which they simulate natural phenomena, but they are never exactly the same as the natural phenomena.

In this activity you will use a model to simulate the greenhouse effect. Materials to use include: a light bulb and receptacle or heat lamp as an energy source; a colorless, transparent plastic container with lid; soil that has been slightly moistened; and, a thermometer.

1. Set the materials up so the light shines on the moistened soil in the enclosed container.
2. Place the thermometer on the soil so that it can easily be read without opening the container. Record the time and temperature every minute for 15 minutes.

3. Data collected by a student and an instructor are recorded in the data table that follows. Graph your data and the data in the table.

Table 5-3   Simulating the Greenhouse Effect Data

Time (minutes)	Temperature (°C) by student	Temperature (°C) by instructor
1	28.0	21.0
2	28.5	21.3
3	29.5	21.6
4	31.0	22.0
5	32.0	23.0
6	32.5	23.3
7	33.0	23.9
8	34.5	24.5
9	35.0	25.0
10	36.0	25.3
11	37.0	25.7
12	38.0	26.0
13	39.0	26.2
14	40.0	26.8
15	41.0	27.5

Concept Development:
4. In comparing this model to the greenhouse effect on Earth, what served in the role of the:
   a. Earth?

   b. atmosphere?

   c. materials in the atmosphere that trap energy?

   d. Sun?

5. Based on your answers to 4a-4d, how does this model differ from the greenhouse effect on Earth?

6. Describe the greenhouse effect on Earth.

7. Account for the differences in temperature recorded by you, the other student, and by the instructor.

Application:
8. Venus has an atmosphere that is primarily made up of carbon dioxide. Briefly explain why Venus is likely to experience the greenhouse effect.

9. Suppose the lid was not used in this activity. Sketch (on the graph you made) how you think the temperature would have changed as measured in the container.

Background Information:
   For additional background, read the **Meteorology** section of the Earth and Space Science Reading.

69　　　Simulating Cloud Formation

Related National Science Education Standards:
  Teaching Standard A: inquiry-based
    Science Content: systems, order, and organization
                                  evidence, models, and explanation
                                  constancy, change, and measurement
                                  evolution and equilibrium
                                  form and function
    Science as Inquiry: abilities to do scientific inquiry
                                  understanding about scientific inquiry
    Physical Science: properties of objects and materials
                                  position and motion of objects
                                  . . . , heat, . . .
                                  properties and changes of properties in matter
                                  motions and forces
                                  transfer of energy
                                  interactions of energy and matter
    Earth and Space Science: properties of Earth materials
                                  objects in the sky
                                  changes in Earth and sky
                                  energy in the Earth system
    Science in Personal and
    Social Perspectives: changes in environments
    History and Nature of
    Science: nature of science

Exploration/Awareness:

Depending on the equipment available, this activity may be done as a lab group activity or as a demonstration by your instructor. Equipment needed includes: large jar, water, matches, and rubber glove.

1. Pour a small amount of water into the large jar. Tip the jar so that water runs on the sides of the jar as well as the bottom of the jar.
2. Strike a match and drop it into the jar so that some of the smoke from the match is "trapped" in the jar.
3. One person needs to place the rubber glove on their hand. Position the gloved hand inside the jar so the top of the glove (the end opposite the fingers) is near the top of the opening of the jar. A second person needs to pull the upper part of the glove so that it is stretched around the outside opening of the jar.

4. If step three has successfully been completed, the person with the glove on should be able to make a fist and slightly push their gloved hand further into the jar (decrease the volume of the air in the jar). Also, they should be able to slightly pull their gloved hand away from the center of the jar and toward the outside of the jar (increase the volume of the air in the jar).
5. Observe and record what happens when the gloved hand is pushed slightly into the jar:

Observation when gloved hand is extended into the jar:

6. Observe and record what happens when the gloved hand in pulled slightly out of the jar:

Observation when gloved hand is pulled away from the center of the jar:

Concept Development:
7. What three components were required to produce a cloud?

8. What role did each of the components serve in the formation of a cloud?

Application:
9. Compare cloud formation to the formation of dew or frost. How are the processes alike and how are they different?

10. Fog is sometimes referred to as a "cloud near the ground." Describe some mechanisms by which fog might form.

Background Information:
For additional background, read the **Meteorology** section of the Earth and Space Science Reading.

70   Relative Humidity

Related National Science Education Standards:
  Teaching Standard A: inquiry-based
    Science Content: systems, order, and organization
                                     evidence, models, and explanation
                                     constancy, change, and measurement
                                     evolution and equilibrium
                                     form and function
    Science as Inquiry: abilities to do scientific inquiry
                                     understanding about scientific inquiry
    Physical Science: properties of objects and materials
                                     . . . , heat, . . .
                                     transfer of energy
                                     conservation of energy . . .
                                     interactions of matter and energy
    Life Science: . . . and environments
                                     matter, energy, and . . . in systems
    Earth and Space Science: properties of Earth materials
                                     objects in the sky
                                     changes in Earth and sky
    Science in Personal and
    Social Perspectives: changes in environments
                                     . . . environments
                                     natural resources
    History and Nature of
    Science: nature of science

Exploration/Awareness:

Relative humidity is a familiar term that most students associate with the study of weather. In addition, when conditions are extremely dry or extremely "muggy," students and others make references to "humidity." In this activity you will study how relative humidity is determined.

1. You will use either a sling psychrometer or construct a device that will function as a psychrometer.
2. If you need to construct the device, use rubber bands or twist ties to attach gauze or cotton balls to the bulb of one thermometer.
3. Tape both thermometers (one with gauze or cotton balls and one without) side by side to a piece of cardboard with the bulbs extending past the edge of the cardboard.

4. Thoroughly wet the gauze or cotton balls on the thermometer by dipping the bulb into some room temperature water. This thermometer will be called the wet bulb thermometer.
5. When the temperature on the wet bulb thermometer stops decreasing, record the temperature in the space provided in step #9.
6. Record the temperature of the dry bulb thermometer in the space provided in step #9.
7. If you are using an already constructed sling psychrometer, you will need to wet one bulb, swing the psychrometer as instructed until the wet bulb temperature stabilizes, and then record both wet and dry bulb temperatures in the space provided in step #9.

Concept Development:
8. Subtract the wet bulb temperature from the dry bulb temperature and record it in the space provided in step #9.
9. Determine the relative humidity using the table below. Locate the dry bulb temperature reading on the left side of the table and the temperature difference (determined in #8; dry bulb temperature – wet bulb temperature) by reading across the top of the table. The relative humidity is read as a % from the table where the dry bulb temperature row intersects the dry bulb – wet bulb temperature column. Example: If the dry bulb temperature is 13 °C and the wet bulb temperature is 8 °C, the relative humidity should be read as 50%.

Dry bulb temperature   _____

Wet bulb temperature   _____

Dry bulb temperature – Wet bulb temperature   _____

Table 5-4   Determining Percent Relative Humidity

Dry Bulb Temperature (°C)	Dry Bulb Temperature – Wet Bulb Temperature (in °C)									
	1	2	3	4	5	6	7	8	9	10
0	81	63	45	28	11					
2	84	68	52	37	22	8				
4	85	70	56	42	29	26	3			
6	86	73	60	47	34	22	14			
8	87	75	63	51	39	28	18	7		
10	88	77	66	55	44	34	24	15	6	
12	89	78	67	57	47	38	29	20	11	3
14	89	79	69	60	51	42	33	25	17	9
16	90	80	71	62	54	45	37	29	22	14
18	91	81	73	64	56	48	41	33	26	19
20	91	82	74	66	58	51	44	37	30	24
22	91	83	75	68	60	53	46	40	34	27
24	92	84	76	69	62	55	49	43	37	31
26	92	85	77	70	64	57	51	45	39	34
28	92	85	78	72	65	59	53	47	42	37
30	93	86	79	73	67	61	55	49	44	39

10. Use Table 5-4 to determine the relative humidity and record it here: _____

11. Typically, the wet bulb temperature is lower than the dry bulb temperature. Explain why.

12. Use the table and by extrapolation determine what happens to the relative humidity when the difference between the wet and dry bulb temperatures becomes increasingly greater.

13. Use the table and by extrapolation determine what happens to the relative humidity when the difference between the wet and dry bulb temperatures becomes increasingly smaller.

14. What do you think the relative humidity is when the wet and dry bulb temperatures are the same? Explain.

15. How is relative humidity determined?

Application:
16. Determine the relative humidity outside. How does it compare to the relative humidity inside?

17. List and describe 2 ways that relative humidity can be changed, i. e., increased or decreased.

Background Information:
For additional background, read the **Meteorology** section of the Earth and Space Science Reading.

71   Dew Point

Related National Science Education Standards:
  Teaching Standard A: inquiry-based
    Science Content: systems, order, and organization
                                      evidence, models, and explanation
                                      constancy, change, and measurement
                                      evolution and equilibrium
                                      form and function
    Science as Inquiry: abilities to do scientific inquiry
                                      understanding about scientific inquiry
    Physical Science: properties of objects and materials
                                      . . . , heat, . . .
                                      transfer of energy
                                      conservation of energy, . . .
                                      interactions of energy and matter
    Life Science: . . . and environments
                                      matter, energy, and . . . in systems
    Earth and Space Science: properties of Earth materials
                                      objects in the sky
                                      structure of the Earth system
                                      changes in Earth and sky
    Science in Personal and
    Social Perspectives: changes in environments
                                      . . . environments
    History and Nature of
    Science: nature of science

Exploration/Awareness:

When discussing weather, meteorologists frequently make reference to the "dew point." Previously, in this course, some other "points" have been studied: boiling point, freezing point, and melting point. In this activity you will determine the dew point.

1. Fill a shiny metal cup about one-third full of room temperature water.
2. Place a thermometer in the cup.
3. Add a **small amount** of ice to the cup and stir with a stirring rod.
4. **Observe very carefully!** Repeat the previous step until moisture or dew appears on the outside of the cup. Record the temperature as soon as the dew appears. This is the dew point temperature.
5. When a good film has appeared, you can "reverse the process" by adding a small amount of warm water and record the temperature when the dew disappears. It is

possible to determine an "average dew point temperature" by averaging the temperature when dew appears with the temperature when the dew disappears.

Concept Development:
6. Define dew point.

7. What is the dew point temperature on the day of your lab?

8. What factors might influence dew point?

9. What happens when the dew point is below 0 °C or 32 °F?

Application:
10. Dew frequently forms on clear, still nights and rarely forms on windy nights. Explain.

11. Describe the relationship, if any, between dew point temperature and relative humidity.

Background Information:
For additional background, read the **Meteorology** section of the Earth and Space Science Reading.

72   Weather Map

Related National Science Education Standards:

Teaching Standard A: inquiry-based	
Science Content:	systems, order, and organization
	evidence, models, and explanation
	constancy, change, and measurement
	evolution and equilibrium
Science as Inquiry:	abilities to do scientific inquiry
	understanding about scientific inquiry
Physical Science:	properties of objects and materials
	position and motion of objects
	properties and changes of properties in matter
	motions and forces
	transfer of energy
	interactions of energy and matter
Earth and Space Science:	properties of Earth materials
	objects in the sky
	changes in Earth and sky
	energy in the Earth system
Science in Personal and Social Perspectives:	changes in environments
	natural hazards
	risks and benefits
History and Nature of Science:	nature of science

Exploration/Awareness:

Weather impacts humans in many obvious ways from the production of crops used for a variety of purposes to severe weather that affects insurance costs and can lead to the loss of lives.

In this activity, you will study how selected weather data is plotted symbolically and used to prepare weather maps. The data include: temperature, dew point, pressure, and wind direction. Recognize that much more data could be used. Once plotted, you will be asked to draw isobars, isotherms, identify areas of high and low pressure, draw fronts, and make weather forecasts.

Data from a weather station are displayed in a station model in a convention agreed upon by meteorologists. What follows is a simplified version of a station model. Temperature in degrees Fahrenheit is recorded in an upper left position and dew point temperature in degrees Fahrenheit is recorded in a lower left position. Pressure, in millibars, is recorded in an upper right position. Average air pressure at sea level is about

1013 millibars (mb). Pressure ranges from about 950 mb to about 1050 mb. When recording pressure on a station model the convention is to write the digits in the tens, ones, and tenths positions with no decimal point. So, 191 represents 1019.1 mb and 890 represents 989.0 mb. A "line segment" from the station model indicates wind direction, i. e., the direction the wind is from. In the example, the wind is from the west. Wind speed is indicated by "flags" attached to the wind direction line segment. A long "flag" represents ten knots and a short one represents five knots. As a unit of speed a knot equals 1.15 statute miles per hour. Shading the station model circle somewhat in proportion to the percentage of clouds indicates the percent of sky covered by clouds.

wind direction = west
wind speed = 15 knots
(long =10; short=5 knots)

temperature   pressure

dewpoint

Concept Development:

Your instructor will provide directions as to how you should proceed to construct a weather map. Data used may depend upon your location and/or the time (season) of the year.

1. Use the weather map and data provided by your instructor to record the weather data at the station models.

2. Follow instructions given to draw isotherms and isobars.

3. Identify areas of high and low pressure and mark them on the map using the symbols H and L.

4. Using temperature data and wind directions, as well as information and symbols provided by your instructor, mark the fronts on the map.

5. Using symbols cP, cT, mP, or mT along with information provided by your instructor, mark the types of air masses on the map.

6. Based on the map constructed and your knowledge of weather, forecast the weather for towns or cities identified by your instructor.

Application:
7. Be prepared to construct a weather map and make weather forecasts.

Background Information:
For additional background, read the **Meteorology** section of the Earth and Space Science Reading.

73   Relative Time

Related National Science Education Standards:
  Teaching Standard A: inquiry-based
  Science Content: systems, order, and organization
                                  evidence, models, and explanation
                                  constancy, change, and measurement
                                  evolution and equilibrium
  Science as Inquiry: abilities to do scientific inquiry
                                  understanding about scientific inquiry
  Physical Science: properties of objects and materials
                                  position and motion of objects
                                  motions and forces
  Earth and Space Science: properties of Earth materials
                                  structure of the Earth system
                                  Earth's history
                                  origin and evolution of the Earth system
  Science in Personal and
  Social Perspectives: changes in environments
  History and Nature of
  Science: science as a human endeavor
                                  nature of science
                                  nature of scientific knowledge
                                  historical perspectives

Exploration/Awareness:

Geologists and other scientists through observations and inferences developed a time scale using relative dating principles. Through careful study, certain sedimentary rock layers, other rock types, and some geologic events were placed in their sequential order of occurrence. Later, technologies developed confirmed much of this sequencing and provided absolute dates. The principles used include:

1. Law of Superposition – In undisturbed sedimentary rocks, each bed or layer is older than the bed above it and younger than the bed below it. The age of rocks becomes progressively younger from bottom to top.
2. Principle of Original Horizontality – Layers of sediment are generally deposited in a horizontal position. Therefore, most sedimentary rock started out with layering close to being horizontal.
3. Principle of Cross-cutting Relationships – A rock or feature must exist before anything like a fault or magma intrusion can intrude into it or cut across it.

4. Principle of Faunal Succession – Fossil organisms succeed one another in recognizable sequences. Sedimentary rocks of different ages contain different fossils and rocks of the same age contain similar fossils. Therefore, any geologic time period can be recognized by its fossil content.

5. In the illustration that follows, use 1 to represent the oldest rock or event, 2 to represent the next oldest rock or event, . . . and 4 to represent the youngest rock or event. Letters A-D represent different sedimentary rock layers.

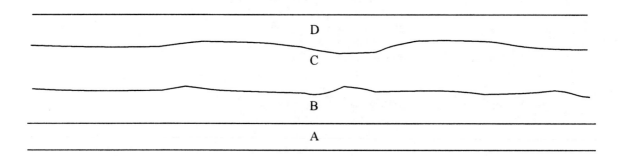

6. In the illustration that follows, use 1 to represent the oldest rock or event, 2 to represent the next oldest rock or event, . . . and 5 to represent the youngest rock or event. Letters A, C, and E represent sedimentary rock layers. Letters B and D represent igneous events.

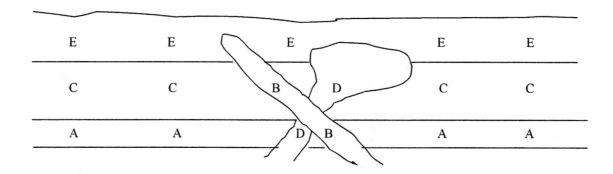

7. In the illustration that follows, use 1 to represent the oldest rock or event, 2 to represent the next oldest rock or event, . . . and 6 to represent the youngest rock or event. Letters A, D, E, and F represent sedimentary rock layers. Letter B represents an igneous event. Letter C represents a fault.

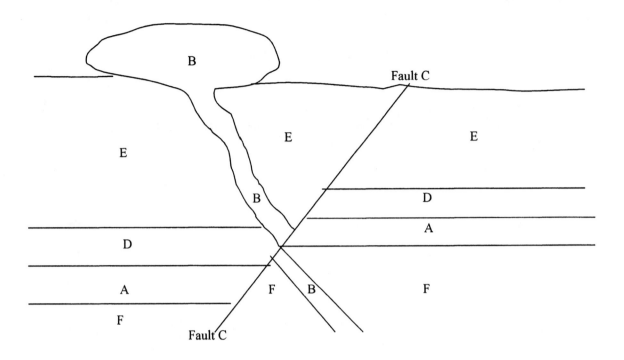

Concept Development:
8. Of the Law of Superposition, Principle of Original Horizontality, Principle of Cross-cutting Relationships, Principle of Faunal Succession, which were used to sequence the sedimentary rock layers illustrated in 5?

9. Of the Law of Superposition, Principle of Original Horizontality, Principle of Cross-cutting Relationships, Principle of Faunal Succession, which were used to sequence the rock layers and geologic events illustrated in 6?

10. Of the Law of Superposition, Principle of Original Horizontality, Principle of Cross-cutting Relationships, Principle of Faunal Succession, which were used to sequence the rock layers and geologic events illustrated in 7?

Application:

11. Use relative dating principles to sequence sedimentary rock layers and geologic events provided as "on paper" practice or in the field as directed by your instructor.

Background Information:

For additional background, read the **Geology** section of the Earth and Space Science Reading.

Name(s) _____

_____

_____

_____

74   Mineral Shapes

Related National Science Education Standards:
  Teaching Standard A: inquiry-based
  Science Content: systems, order, and organization
                              evidence, models, and explanation
                              constancy, change, and measurement
                              form and function
  Science as Inquiry: abilities to do scientific inquiry
                              understanding about scientific inquiry
  Physical Science: properties of objects and materials
                              . . . , magnetism
                              properties and changes of properties in matter
                              structure and properties of matter
                              interactions of energy and matter
  Earth and Space Science: properties of Earth materials
                              geochemical cycles
  Science and Technology: . . . distinguish between natural objects . . .
  Science in Personal and
  Social Perspectives: types of resources
                              . . . resources, and environments
                              natural resources
  History and Nature of
  Science: science as a human endeavor
                              nature of science

Exploration/Awareness:

Earth scientists study rocks and minerals because they are the materials that make up the solid Earth. Minerals, along with other materials that are mined, combine with agricultural products to form the raw materials that societies use. Individuals' lives, relationships between countries, and economies are affected by the availability of minerals and their trade. Minerals are commonly described as: naturally occurring (not human-made), inorganic ("non-living" or lacking carbon-hydrogen bonds), solid, with a definite chemical composition and crystalline structure.

Various properties can be used to identify minerals. Variations in specimens or impurities in specimens can be problematic when identifying minerals. For these reasons, it is important to look at a variety of samples. Chemical composition and crystal structure distinguish each mineral from all others. All minerals have atoms arranged in a regular, periodically repeated pattern. This arrangement is the crystalline structure of a mineral. Crystals that grow freely in an uncrowded environment develop a crystal face, which is a flat surface. Other simultaneously growing crystals or previously formed minerals frequently impede the growth of later forming crystals in nature. For this reason, minerals rarely show perfect development of crystal faces. Minerals that show perfect development of crystal faces are expensive to buy.

In this activity you will closely examine some artificial mineral shapes and see if you can determine a relationship among the number of faces, edges, and vertices ("points").

1. Use the shapes provided by your instructor or construct the shapes. Examine the cube, the tetrahedron, and octahedron. Complete the table that follows by counting and recording the number of faces, edges, and vertices.

Shape	Number of Faces	Number of Vertices	Number of Edges	
cube				
tetrahedron				
octahedron				

Concept Development:
2. Look at the number of faces, number of vertices, and the number of edges for each of the shapes. Do you see a relationship? If you do, describe the relationship. If you do not, title the last column "Faces + Vertices" and add the number of faces and the number of vertices for each shape. Now describe the relationship.

3. From your previous study of minerals you might recognize pyrite ("fool's gold") and halite ("salt"). Each of these minerals may have crystals in the shape of a cube. How might they be distinguished from one another?

Application:
4. Test the relationship you developed for #2 with other regular three-dimensional crystal shapes. Does the relationship hold true?

5. Use a reference to look up a scientist with the last name of Euler. How did he describe the relationship you developed in #2?

Background Information:
    For additional background, read the **Minerals** section of the Earth and Space Science Reading.

Name(s)

_____

_____

_____

_____

75   **Rock Identification**

Related National Science Education Standards:
  Teaching Standard: inquiry-based
    Science Content: systems, order, and organization
                                  evidence, models, and explanation
                                  constancy, change, and measurement
                                  evolution and equilibrium
    Science as Inquiry: abilities to do scientific inquiry
                                  understanding about scientific inquiry
    Physical Science: properties of objects and materials
                                  properties and changes of properties in matter
                                  structure and properties of matter
                                  interactions of energy and matter
    Earth and Space Science: properties of Earth materials
                                  Earth's history
                                  geochemical cycles
    Science in Personal and
    Social Perspectives: types of resources
                                  . . . resources . . .
    History and Nature of
    Science: nature of science

Exploration/Awareness:

     Rock types can be identified by their textures and composition. Texture refers to
the size, shape, and boundary relations between adjoining minerals in a rock.
Composition refers to the minerals or elements that make up the rock.

1. Observe the numbered sedimentary rock specimens provided by your instructor. Use
    the texture terms that follow to help add information to Table 5-5. Regarding
    composition, if you can identify minerals that make up the rock, do so. If you cannot
    identify individual minerals, identify the rock as being composed of principally one
    mineral or more than one mineral. In the miscellaneous column, record any
    observations that are distinctive that would help you identify the rock type.

    Texture terms: clastic, crystalline, oolitic, skeletal

2. Observe the numbered igneous rock specimens provided by your instructor. Use the
    texture terms that follow to help add information to Table 5-6. Regarding
    composition, if you can identify minerals that make up the rock, do so. If you cannot
    identify individual minerals, identify the rock as being composed of principally one

mineral or more than one mineral. In the miscellaneous column, record any observations that are distinctive that would help you identify the rock type.

Texture terms: phaneritic, aphanitic, glassy, porphyritic-phaneritic, porphyritic-aphanitic, fragmental

Table 5-5   Sedimentary Rock Observations

Rock Type	Texture	Composition	Miscellaneous
conglomerate (1)			
breccia (2)			
sandstone (3 & 4)			
arkose			
graywacke			
siltstone			
shale (5)			
crystalline limestone (sparite 8?)			
microcrystalline limestone (8?)			
oolitic limestone			
coquina			
fossiliferous limestone			
chalk (7)			
travertine			
dolomite			
chert or flint (6)			
(rock) gypsum			
(rock) salt			

Table 5-6   Igneous Rock Observations

Rock Type	Texture	Composition	Miscellaneous
granite (14, 15)			
rhyolite			
diorite			
andesite			
gabbro			
basalt (16)			
peridotite			
obsidian (17)			
pumice or scoria (18)			
tuff			
volcanic breccia			
aggloomerate			

3. Observe the numbered metamorphic rock specimens provided by your instructor. Use the texture terms that follow to add information to Table 5-7. Regarding composition, if you can identify minerals that make up the rock, do so. If you cannot identify individual minerals, identify the rock as being composed of principally one

mineral or more than one mineral. In the miscellaneous column, record any observations that are distinctive that would help you identify the rock type.

Texture terms: foliated, nonfoliated

Table 5-7   Metamorphic Rock Observations

Rock type	Texture	Composition	Miscellaneous
slate (11)			
phyllite			
schist (10)			
gneiss (9)			
metaconglomerate			
quartzite (13)			
marble (12)			

Concept Development:
4. What features help distinguish igneous rocks from sedimentary rocks from metamorphic rocks?

5. What features allow you to distinguish among specific igneous rock types?

6. What features allow you to distinguish among specific sedimentary rock types?

7. What features allow you to distinguish among specific metamorphic rock types?

Application:
8. Study rock types identified by your instructor. Be prepared to identify rock types by sight on the unit test.

Background Information:
      For additional background, read the **Geology** section of the Earth and Space Science Reading.

76	**Soil**

Related National Science Education Standards:	
Science Content:	systems, order, and organization
	evidence, models, and explanation
	constancy, change, and measurement
	evolution and equilibrium
Science as Inquiry:	abilities to do scientific inquiry
	understanding about scientific inquiry
Physical Science:	position and motion of objects
	motions and forces
	light, . . .
Earth and Space Science:	objects in the sky
	changes in Earth and sky
	structure of the Earth system
	energy in the Earth system
Science in Personal and Social Perspectives:	changes in environments
History and Nature of Science:	science as a human endeavor
	nature of science
	nature of scientific knowledge
	historical perspectives

Exploration/Awareness:

Carefully observe soil samples provided only by your instructor. Some soil samples may not be safe to handle.

1. Observe the soil samples. Record your observations.

2. If the information is available, note where the soils were obtained, i. e., their source and location.

Concept Development:
3. What is in soil, i. e., what is soil composed of?

4. Compare the size of particles that make up soils.

5. How are soils different?

Application:
6. Design and conduct a control variable experiment using soils. Consider the following as variables that might be investigated:
    a. size of soil particles.
    b. ability to "soak up" water.
    c. flow rates of water through soil.
    d. factors that affect soil erosion.
    e. "best" soil for plant growth.
    f. settling rate of soil particles in water.

Check with your instructor before proceeding with your soil control variable experiment.

Background Information:
For additional background, read the **Geology** section of the Earth and Space Science Reading.

77   **Studying Fossils at a Geology Museum**

Related National Science Education Standards:
Teaching Standard A: inquiry-base
Science Content: systems, order, and organization
                                              evidence, models, and explanation
                                              constancy, change, and measurement
                                              evolution and equilibrium
                                              form and function
Science as Inquiry: abilities to do scientific inquiry
                                              understanding about scientific inquiry
Earth and Space Science: properties of Earth materials
                                              Earth's history
                                              origin and evolution of the Earth system
Science in Personal and
Social Perspectives: changes in environments
                                              populations, resources, and environments
History and Nature of
Science: nature of science
                                              nature of scientific knowledge
                                              historical perspectives

Exploration/Awareness:
        Use directions provided by your instructor to complete Table 5-8. Use information available in the museum to answer questions and complete tasks.
1. Complete the table by identifying examples of the fossil, identifying a specific example or common name, and then making a generalized sketch. Identify the fossil as a plant (Pl), animal (An), fungi (F), protist (Pr), archaebacteria (Ar), or eubacteria (Eu). Describe its habitat if available, and then identify the time in which the fossil lived.

Concept Development:
2. What is a fossil?

3. Identify some processes by which fossils might form.

4. What is the Geologic Time Scale?

Application:

5. Make inferences about past environments based on fossils found near where you live.

6. Your instructor will identify fossils that you will be asked to identify on the exam. Review these prior to the exam.

Table 5-8   Fossils

Fossil	Specific Example and/or Common Name	Sketch	Pl, An, F, Pr, Ar, Eu?	Environment or Habitat	Time
Algae					
Angiosperm					
Arthropoda					
Dinosaur (any type)					

Fossil	Specific Example and/or Common Name	Sketch	Pl, An, F, Pr, Ar, Eu?	Environment or Habitat	Time
(Echinodermata) Asteroida					
(Echinodermata) Blastoidea					
Brachiopoda					
Bryozoa					
Calamites					
(Mollusca) Cephalopoda					
(Coelenterata) Corals & related					
Cordaites					

Fossil	Specific Example and/or Common Name	Sketch	Pl, An, F, Pr, Ar, Eu?	Environment or Habitat	Time
(Echinodermata) Crinoidea					
(Echinodermata) Cystoidea					
Diatoms					
(Echinodermata) Echinodea					
Fern					
Fish					
(Protozoa) Foraminifera					
(Protozoa) Foraminifera Fusilinid					

Fossil	Specific Example and/or Common Name	Sketch	Pl, An, F, Pr, Ar, Eu?	Environment or Habitat	Time
(Mollusca) Gastropoda					
Graptolites					
Leaf					
Lepidoden-dron					
Mammoth					
Mastodon					
Mosasaur					
(Mollusca) Pelecypoda					

Fossil	Specific Example and/or Common Name	Sketch	Pl, An, F, Pr, Ar, Eu?	Environment or Habitat	Time
Petrified Wood					
Protozoa					
Pteranodon					
(Protozoa) Radiolaria					
Receptacu-lites					
Saber-toothed cat					
Sigillaria					
Sloth					

Fossil	Specific Example and/or Common Name	Sketch	Pl, An, F, Pr, Ar, Eu?	Environment or Habitat	Time
(Porifera) Sponges and sponge-like					
Turtle					

Background Information:

For additional background, read the **Geology** section of the Earth and Space Science Reading.

78   **Plate Tectonics Illustrations and Explanations**

Related National Science Education Standards:	
Science Content:	systems, order, and organization
	evidence, models, and explanation
	constancy, change, and measurement
	evolution and equilibrium
Science as Inquiry:	abilities to do scientific inquiry
	understanding about scientific inquiry
Physical Science:	position and motion of objects
	motions and forces
	light, . . .
Earth and Space Science:	objects in the sky
	changes in Earth and sky
	structure of the Earth system
	energy in the Earth system
Science in Personal and Social Perspectives:	changes in environments
History and Nature of Science:	science as a human endeavor
	nature of science
	nature of scientific knowledge
	historical perspectives

Information will be presented via lecture, video, or computer software that illustrates different types of plate boundaries, examples, and associated landforms or features. Draw illustrations and take notes.

Divergent oceanic crust
Examples
Landforms/Features

Divergent continental crust
Examples
Landforms/Features

Convergent oceanic crust
Examples
Landforms/Features

Convergent oceanic-continental crust
Examples
Landforms/Features

Convergent continental crust
Examples
Landforms/Features

Transform oceanic crust
Examples
Landforms/Features

Transform continental crust
Examples
Landforms/Features

79   **Geoscience Maps**

Related National Science Education Standards:
  Teaching Standard A: inquiry-based
  Science Content: systems, order, and organization
                                      evidence, models, and explanation
                                      constancy, change, and measurement
                                      evolution and equilibrium
                                      form and function
    Science as Inquiry: abilities to do scientific inquiry
                                      understanding about scientific inquiry
    Physical Science: properties of objects and materials
                                      position and motion of objects
                                      properties and changes of properties in matter
                                      motions and forces
                                      transfer of energy
                                      interactions of energy and matter
    Earth and Space Science: properties of Earth materials
                                      changes in Earth and sky
                                      structure of the Earth system
                                      Earth's history
                                      energy in the Earth system
                                      . . . evolution of the Earth system
  Science in Personal and
  Social Perspectives: types of resources
                                      changes in environments
                                      natural hazards
                                      risks and benefits
                                      science and technology in society
                                      natural resources
                                      natural and human-induced hazards
  History and Nature of
  Science: science as a human endeavor
                                       nature of science
                                       history of science
                                       nature of scientific knowledge
                                       historical perspectives

Exploration/Awareness:

Humans make use of maps in many obvious ways. Maps in the geosciences are used to visualize and make inferences about the relative ages of rocks, identify structural features, and investigate deformation of the Earth's crust (tectonic activity).

In this activity, you will have the opportunity to use both topographic and geologic maps. Depending on the time available and your instructor's wishes, the maps you study may be limited to maps of your local area or maps from other areas may be used to illustrate a greater variety of geoscience phenomena.

Follow directions provided by your instructor to identify items to be studied from the various maps provided. For geologic maps, these may include but are not limited to:

anticlines	basin	batholiths
compressional fault	dikes	dip
dome	epochs	eras
faults	folds	formations
groups	nonplunging	periods
plunging	sills	strike
synclines	systems	tensional fault

For topographic maps, items to be studied may include but are not limited to:

base line	bench mark (B. M.)	cirque
contour interval	contour lines	degree
depression	elevation	glacial trough
ground moraine	kettle	latitude
longitude	map scale	mature region
meander	meander cut off	meridians
minute	moraine	old age region
outwash plain	oxbow lake	parallels
principal meridian	quadrangles	range lines
relief	second	section
sink hole	spot elevation	terminal moraine
township lines	U-shaped valley	V-shaped valley
youthful region		

Concept Development:

Your instructor will provide directions as to how you should proceed, what maps will be used, and what topics will be emphasized.

Application:

Be able to use both geologic and topographic maps.

Background Information:

For additional background, read the **Geology** section of the Earth and Space Science Reading.

80   **Investigating Earth Science Hazards**

Related National Science Education Standards:	
Science Content:	systems, order, and organization
	evidence, models, and explanation
	constancy, change, and measurement
	evolution and equilibrium
Science as Inquiry:	abilities to do scientific inquiry
	understanding about scientific inquiry
Physical Science:	position and motion of objects
	motions and forces
	light, . . .
Earth and Space Science:	objects in the sky
	changes in Earth and sky
	structure of the Earth system
	energy in the Earth system
Science in Personal and Social Perspectives:	changes in environments
History and Nature of Science:	science as a human endeavor
	nature of science
	nature of scientific knowledge
	historical perspectives

Floods, earthquakes, volcanic eruptions, mass wasting, tornadoes, hurricanes, and other forms of severe weather are examples of earth science hazards. Select one of the hazards listed or identify another type of hazard. Collect data on the:

1. science behind the hazard, i. e., describe what causes the hazard to occur.
2. frequency of the hazard.
3. damage caused by the hazard.
4. locations where the hazard occurs and make a map illustrating frequencies.

5. Properly cite any references used.

81   **Earth and Space Science-Related Hobby Report**

Significant numbers of people engage in earth/space science-related activities outside their time at work. They pursue these activities as hobbies out of curiosity, for the enjoyment it brings, and for the benefits they bring to their lives.

What follows is a list of some hobbies or activities where possession of some earth/space science knowledge and/or skills would be useful. Other possibilities exist.

fossil collecting   jewelry making/rock tumbling
mineral collecting   rock collecting
storm spotter

Select one of the possibilities from the list or come up with a earth/space science-related hobby or activity of your own (get your instructor's approval on your idea). Write a 1.5 to 2 page, double-spaced report on the hobby or activity. Address the following in your report:
1. Identify the earth/space science knowledge and skills needed to participate in this hobby or activity. This must be included in the report.
2. Describe the role, if any, that technology plays in this hobby or activity. This must be included in the report.
3. Report the approximate number of people that participate in this hobby or activity. Include this information as available.
4. If references are used, they must be properly cited.

Name(s)

_____

_____

_____

_____

82   Scientists, Nature of Science, and History of Science Report

Scientists working in the earth and space sciences have made significant contributions to human history. Select an important earth/space science concept, idea, or theory and identify scientists that contributed to the development of the concept, idea, or theory. Write a 2 to 3 page, double-spaced report on the scientists and topic selected. Address the following in your report:

1. Clearly identify the earth/space science concept, idea, or theory selected. This must be included in the report.
2. Describe the role that scientists played in the development of this concept, idea, or theory. This must be included in the report.
3. Describe how the historical development of the concept, idea, or theory illustrates the nature of science. This must be included in the report.
4. If references are used, they must be properly cited.

83   **Earth and Space Science Questions and Problems**

Note: To receive full credit on problems, you must show your work.

1. Compare characteristics used to identify minerals with those used to identify rocks.

2. If a scale of 1 cm = 5,000,000 years is used, determine the scale distance that would represent the Jurassic period.

3. Make a sketch that illustrates the location of a first or waxing quarter Moon at 7:00 PM.

4. Use temperatures to describe a set of conditions that would produce a relative humidity of 75%.

5. Which type of plate boundaries:
    a. are least likely to form volcanoes?

    b. produce earthquakes?

6. Identify and briefly describe at least three differences between cold fronts and warm fronts and "typical weather" associated with each.

7. Compare stars according to their temperatures, sizes, and what they are made of.

8. What is necessary for precipitation to form?

9. Describe factors that contribute to the form or shape of the three main cloud types.

10. Briefly describe the principles or ideas involved for scientists to determine the relative sequence of geological events.

11. What features can be used to identify:
    a. brachiopods?

    b. crinoids?

    c. petrified wood?

84   Earth and Space Science Reading

(Note: See other Activity-Based Physical Science units for additional reading on selected physical science topics.)

Introduction

Earth science or earth system science is a study of the Earth's interacting or interdependent parts. The parts are connected so that changes in one part can produce changes in other parts. Some references refer to these parts as spheres. The spheres are identified as: hydrosphere (water), atmosphere (gaseous mass around Earth), biosphere (regions of Earth that support life), and lithosphere (solid Earth to some but a portion of the solid Earth to others). Depending on the source, earth science includes the subdisciplines of: **astronomy, geology, hydrology, oceanography, meteorology, climatology,** and **paleontology**.

Astronomy

The motion of one celestial body around another is called **revolution**. The time it takes for Earth to complete one revolution in an elliptical path around the Sun is a year. The path of a body in revolution around a center of mass is called its **orbit**. The spinning of a celestial body on its axis is called **rotation**. The time it takes the Earth to rotate once on its axis is a **day**.

Our **solar system** includes the **Sun**, **nine planets**, their **moons**, and **asteroids**. The inner planets, **Mercury**, **Venus**, **Earth**, and **Mars** are rocky or terrestrial. The outer planets, **Jupiter**, **Saturn**, **Uranus**, **Neptune**, and **Pluto** are very different from the inner planets. Other than Pluto, they have much larger diameters. The large outer planets can almost be thought of as balls of gas. Even though they all have small cores of rock and iron, they are mostly made of gases. Pluto is unique and does not readily lend itself to being classified as similar to one of the inner, terrestrial planets or outer, gaseous planets. There have been debates as to whether Pluto qualifies as a planet. See the Scale Model of the solar system activity for more numeric information on the planets.

The Sun is our nearest star. Solar energy streams out from the Sun in all directions. The Sun is the primary source of energy for Earth even though Earth receives about one two-billionths of the Sun's total energy. Most of the information scientists have about the universe is obtained from the study of light emitted from celestial bodies. Humans are most sensitive to visible light. Visible light is a small portion of **electromagnetic radiation**. Electromagnetic radiation includes (from short waves to longer waves): cosmic rays, gamma rays, X-rays, ultraviolet light, visible light (violet, indigo, blue, green, yellow, orange, red), infrared radiation, radar and microwaves, FM radio and television, short wave, AM radio, and long radio waves. The Sun is an average star in terms of size and temperature.

Solar energy from the Sun consists mainly of light and heat that travel through space. The energy is derived from nuclear **fusion**; energy formed when the nuclei of atoms join together. The Sun is more than a trillion times the volume of Earth and about 333,420 times the Earth's mass. Gigantic pressure in the Sun's center allows the fusing together of hydrogen nuclei. The energy released produces temperatures of about 27 million degrees Fahrenheit of about 15 million degrees Celsius in the Sun's core.

The photosphere appears as the bright disk of the Sun because it radiates most of the sunlight observed. The photosphere is considered the surface of the Sun although it is made up of a layer of incandescent gas. The photosphere glows at more than 9900 °F or about 5500 °C. The Sun's surface has bright spots called granules. **Sunspots** are dark regions on the surface that appear darker because they are cooler. Sunspots are associated with intense magnetic disturbances. Above the photosphere is the Sun's atmosphere with the chromosphere below the corona. Solar prominences are flamelike tongues of hot hydrogen that shoot out 100,000 km or about 60,000 miles.

Mercury is the closest planet to the Sun at a distance of about 36 million miles. It is the second smallest planet in the solar system. The planet's proximity to the Sun contributes to its wide temperature range. The illuminated portion of the planet may reach a high temperature of about 870 °F to a low of about –300 °F on the side opposite the Sun. A year on Mercury is equal to about 88 Earth days. Analysis of pictures taken by space probes reveal the surface of Mercury to have been affected by volcanic eruptions and impact craters.

Venus is covered with clouds and has an atmosphere that traps infrared radiation pushing the temperature to between 850 °F and 900 °F making it the hottest planet. Incoming solar radiation heats the surface of Venus. Venus' atmospheric gases, in what is known as the "greenhouse effect," trap infrared radiation coming from the ground. Venus rotates slowly on its axis; it takes about 243 Earth days for Venus to complete one rotation. Its rotation is retrograde, i. e., it rotates clockwise when viewed from above. Other planets rotate in a counterclockwise direction when viewed from above. Because of its thick cloud cover, astronomers have used radar to determine that the surface of Venus has many volcanoes. As technology improves, astronomers are able to find out more about Venus' atmosphere and surface.

Earth revolves around the Sun at an average distance of about 93 million miles. Earth is the largest of the terrestrial planets with a diameter of about 7926 miles. Earth's natural satellite, the Moon, is the largest satellite of the inner planets. While Mercury, Venus, and Mars show evidence of impact craters, surface processes like erosion and the creation and destruction of crustal material have obscured this aspect of Earth's geological history. The average temperature on Earth is about 45 °F. Temperatures on Earth allow water to exist in solid, liquid, and gaseous states.

Earth's **Moon** orbits the Earth in an elliptical path that averages about 237,000 miles (SI range from 356,000 km to 407,000 km). One theory suggests the Moon formed after a Mars-sized object collided with Earth. U. S. astronauts have made the Moon the only astronomical body that humans have stood on besides Earth. It takes 27.3 days for the Moon to orbit Earth with respect to the background of the stars (hardly changes). This is a sidereal month. It takes the Moon about 29.5 days to orbit the Earth with respect to the Sun (because the Earth has moved in its own orbit). This is a synodic month. Always half the Moon reflects light from the Sun. The sunlit portion of the Moon that we see from Earth produces the **phases of the Moon**. The phase we see from Earth is dependent upon the Sun-Earth-Moon angle. New Moon occurs when the Moon is exactly in line between the Sun and Earth (Sun-Earth-Moon angle is 0°). A full Moon occurs when the Earth is exactly in line between the Sun and Moon (Sun-Earth-Moon angle is 180°). Quarter Moon occurs when the Sun-Earth-Moon angle is 90°. A crescent

Moon occurs when the Sun-Earth-Moon angle is between 0° and 90°. A gibbous Moon occurs when the Sun-Earth-Moon angle is between 90° and 180°.

A **solar eclipse** occurs when the Moon is at new phase and the Moon covers the Sun's disk. A **lunar eclipse** occurs during full Moon when the Moon enters the shadow of the Earth. The Moon's plane of orbit is tilted slightly from that of the Sun-Earth plane, causing eclipses to occur less frequently than if they were all in the same plane.

Mars is a good first candidate for interplanetary travel because it is nearest to the Earth and also because the Martian environment seems to be most like Earth's. A Mars day is similar in length to an Earth day. Scientists have made attempts to determine if life ever existed on Mars and have searched for water on Mars. As this is written some NASA scientists have evidence that leads them to infer that life may have once existed on Mars. Mars has a reddish tint when observed with the naked eye and is known as the "Red Planet." Mars possesses the solar system's largest volcano, Mons Olympus, at about 75,000 feet high. Mars has two small moons.

With a diameter of nearly 89,000 miles, **Jupiter** is the largest planet in the solar system. Its atmosphere is mostly hydrogen. Some have suggested that if Jupiter was about ten times larger it would have evolved into a star. Today, 17 of Jupiter's moons have been observed. Using a telescope, Galileo first observed Jupiter's four largest moons. The two largest, Europa and Io, are about the size of Earth's moon. Along with Saturn, Jupiter is a favorite for planetary observers because its large moons can be observed with a telescope. Jupiter has a faint ring system. Jupiter is famous for its Great Red Spot. The Great Red Spot is a counterclockwise rotating storm. Jupiter has an atmosphere made up of mostly hydrogen and changes appearance daily. Jupiter has huge storms and pictures taken illustrate cloud features and lightning.

Saturn is famous for its most prominent feature, its rings. The ring system is only tens of yards thick but is about 169,800 miles in diameter. Saturn's rings make it a favorite for planet observers. Using a telescope, Galileo discovered the rings in 1610. His telescope could only resolve the ring system into what appeared to be two smaller bodies. Fifty years later, Christian Huygens identified the ring nature of what Galileo first observed. It is thought the rings are made up of small particles ranging in size from about ten centimeters to around ten meters. For many years, Saturn was thought to be the only planet with a ring system. Jupiter, Uranus, and Neptune all have faint ring systems. Saturn has many moons and the largest, Titan, is the second largest (to Jupiter's Ganymede) moon in the solar system. Titan is thought to have an atmosphere of about 80% nitrogen and possibly 6% methane.

Uranus rotates on its side. It is inferred that this could be the result of a collision with a large object. The rotation of other planets is characterized as "spinning like tops." In contrast, Uranus might be considered as almost "rolling like a ball."

Because of its distance from the Sun, **Neptune** is cold with a temperature of about –370 °F. Neptune has storms in its atmosphere and technology has detected the presence of a massive cloud and smaller clouds. Triton, its largest moon, orbits Neptune clockwise. This is opposite the orbits of the planets around the Sun and the major moons around their planets. Because of this, some have inferred that Triton formed independent of Neptune and was captured by the planet's gravitational attraction.

With a diameter of about 1485 miles, **Pluto** is the smallest planet in the solar system. Pluto's characteristics do not allow it to be classified as a terrestrial planet nor as one of the gas giants. There has been debate about whether Pluto should be considered a planet; Pluto may be just one of many icy objects that orbit the Sun at great distances. Pluto's moon, Charon, is close to the size and brightness of Pluto. In contrast to some that question whether Pluto is a planet, some astronomers think of Pluto and Charon as a double planet. Pluto's elliptical orbit places it in a range of about 2.8 billion miles to 4.6 billion miles from the Sun. About every 250 years, Pluto's orbit takes it inside the orbit of Neptune for about 20 years at a time.

In addition to the planets and their moons, thousands of smaller bits of rock and ice orbit the Sun. They range in size from small dust particles to being hundreds of miles across. **Asteroids** are thought to be leftover fragments from the formation of the solar system too scattered to coalesce to form a planet. Ceres, the largest asteroid, has about a 600 mile diameter. Most observed asteroids are only one kilometer in diameter. The smallest asteroids are assumed to be sand grain sized. Most asteroids lie in a wide band between Mars and Jupiter but some come close to Earth and in the geologic past have occasionally crashed into it. A **meteoroid** is a small solid body. Once a meteoroid enters Earth's atmosphere it is called a **meteor** and friction with the Earth's atmosphere may cause it to glow. This is the phenomenon that many refer to as a "shooting star." A meteoroid that does not completely burn up in the atmosphere and hits the ground is called a **meteorite**. Micrometeorites are so small that their rate of fall is too slow to cause them to burn up so they fall as space dust. Meteorites are classified by their composition: 1) irons – mostly iron with some nickel, 2) stony – silicate minerals with some other minerals, and 3) stony-irons – which are mixtures. **Comets** are frequently described as "dirty snowballs." They are made of frozen gases, including water, which hold together small pieces of rocky and metallic materials. The central part of a comet may be thought of as an irregular shaped chunk of ice covered with dust. As a comet gets closer to the Sun it starts to melt from the heat. Dust, vapors, and gases form a tail that can be millions of miles long blown by the solar wind in a direction that faces away from the Sun. When the comet passes by the Sun and heads back out into deep space, the solar wind causes the tail to face the direction it is moving. Comets follow long elliptical orbits around the Sun. From deep out in space, they swing around the Sun and then head out into space to return years or centuries later. If Earth's orbit takes it through a comet's tail, the particles that burn up in Earth's atmosphere are called meteors and may be viewed as "meteor showers."

Larger chunks of "space rock" may crash into a planet or moon producing a **crater**. Traveling at high speeds, huge holes can be formed at the surface of planets or moons. Our Moon's surface is covered with meteorite craters. Meteor Crater near Winslow, Arizona is a famous example that is nearly three-fourths of a mile wide and about 560 feet deep.

Stars are familiar objects observed in the night sky. Stars can be classified by their color, brightness, mass, temperature, and size. **Magnitude** is a number given to a celestial object to express its relative brightness. **Apparent magnitude** is the brightness of a star when viewed from Earth. **Absolute magnitude** refers to the brightness of a star if it were viewed from a distance of 10 parsecs (32.6 light years). This is used to compare the true brightness of stars. Each difference of one in magnitude is a difference of 2.512 times in brightness. So, a star of the first magnitude is about 2.5 times brighter than a star of the second magnitude. Most stars have absolute magnitudes in the range between –5 and 15. For example, the Sun has an absolute magnitude of 5 and Sirius has an absolute magnitude of 1.5. The Sun has an apparent magnitude of –26.7, Venus –4.3, and Sirius –1.4. For convenience in referring to positions in the sky, astronomers have grouped stars into 88 **constellations**. Constellations are apparent groupings; although the stars of a constellation look close together in space they may actually be separated by large distances and vary greatly in their distances from Earth. Constellations have been named for mythical characters. **Galaxies** represent large volumes of space containing billions of stars held together by their mutual gravitational attraction. Our own **Milky Way Galaxy** is a spiral galaxy. Spiral galaxies are rotating galaxies that have flattened shapes with "arms" that radiate out from the center like a pinwheel.

Star formation begins when clouds of **gas** and **dust** called **nebulae** start to coalesce. Where the gas and dust is more dense (more mass per unit of space), gravity pulls "clumps" together. As the material contracts, temperature and pressure increase. The temperature increases but perhaps is not hot enough to produce a glow. In this situation smaller clumps (less gravity) may fizzle out as brown dwarves. Larger clumps may reach a temperature of around 10,000,000 °C (about 18,000,000 °F). Near this temperature nuclear **fusion**, the joining of atomic nuclei, occurs. Nuclear fission, the splitting of nuclei, is the energy producing process used by nuclear reactors. Mass is converted into energy. The famous equation, $E = mc^2$, calculates the energy produced from the conversion of mass. E, energy, equal the mass times the speed of light squared. The speed of light is about 186,000 miles per second. Nuclear fusion is an example of nuclear change where new **elements** are formed. At the core of a star, **hydrogen** is fused to form **helium**. Helium can then serve as the fuel. Nuclear fusion produces energy that causes a star to shine. Stars are stable when the heat produced by fusion produces a pressure or "outward force" that balances the "inward force" due to **gravity**. Stable stars can stay about the same size and glow for billions of years. When the hydrogen is used up, the force of gravity may cause the core of the star to shrink and begin to fuse helium. The gaseous outer layers may expand and the star can become a **red giant**. In more massive stars, more changes can take place in the core. Iron may be made in the core. The star takes in energy instead of releasing energy. The star may collapse suddenly, within seconds. The collapse of more massive stars produces a giant explosion called a **supernova**. Supernovae shine with the brightness of millions of suns for a few weeks. The remains of a supernova may form a rapidly spinning star made of mostly neutrons called a **pulsar**. A pulsar is a neutron star that emits a pulsating radio signal. A **neutron star** is a star of very high density composed entirely of neutrons. Stars smaller than our Sun gradually cool and shrink into a **white dwarf**. A white dwarf represents a stage in the development of a star where fusion has stopped and the star glows as a result of leftover heat. As the name implies, a white dwarf is a small star. A **black hole** forms when a massive star collapses to a small volume creating matter packed so densely that light cannot escape from its gravitational field.

Independently, Ejnar Hertzsprung and Henry Russell plotted star brightness along a vertical axis and spectral class or temperature along a horizontal axis. The **Hertzsprung-Russell (H-R) diagram** is called by some the most important diagram in astronomy. It shows a pattern in the size and temperature of stars. The **main sequence** is the most prominent region of the H-R diagram. The more luminous stars on the main sequence are hot and bluish white while the dimmest stars are cool and reddish. Our Sun falls somewhere in the middle of these extremes. A star spends most of its existence on the main sequence. Hydrogen is converted into helium. When the hydrogen is used up, the star moves off the main sequence as a red giant. At this stage the star enters the "helium burning" stage of development. Nuclear reactions involving helium take place. One of these involves the formation of a heavier element, **carbon**. The Sun is about halfway through its existence as a main sequence star. In approximately 5 billion years, it is thought the Sun will evolve into a red giant. Once stars evolve into the helium burning stage, they change by taking one of three directions. They either become white dwarfs, neutron stars, or black holes. At this time, stars have used up nearly all of their nuclear fuel. Stars about 1.2 solar masses or lighter can end their existence as white dwarfs. They are relatively hot but small stars. Neutron stars have an internal pressure so great that plasma electrons combine with protons to form neutrons. Neutron stars are only a few kilometers in diameter and spin rapidly. Pulsars are rapidly rotating neutron stars. It is thought that extremely massive stars may collapse to densities greater than those found in neutron stars. Gravitational fields around such stars, black holes, would be so great as to prevent matter or light from leaving them.

The **Big Bang Theory** proposes that 10 to 20 billion years ago the universe originated when all matter exploded from a single, compressed mass. This event is thought to mark the beginning of our universe.

Earlier in the Activity-Based Physical Science reading you had the opportunity to read about locating objects in the sky using one of two systems. The **altitude-azimuth** system has the advantage of being easy to use. This system is useful in the vicinity of the observer's location but it is a disadvantage that altitude-azimuth measurements vary based on observers being in different geographic locations. Altitude in this context refers to height above the horizon measured in degrees. The point directly overhead, i. e., at 90°, is called the **zenith**. Altitude can range from 0° (at the horizon) to 90°. Azimuth refers to a horizontal angular distance from a fixed reference point. North will be described as being either 0° or 360°, east at 90°, south at 180°, west at 270°, and points in-between are represented by values in-between these numbers. For example, southwest is represented by 225°.

The second system uses **right ascension** and **declination**. This system has the advantage of being universal, i. e., it is the same for all observer locations on Earth, but has the disadvantage of being more difficult to use. The right ascension/declination system is analogous to the latitude/longitude system for describing locations on Earth. Declination is the celestial equivalent of latitude and is measured in degrees north (+ or positive) or south (– or negative) of the extension of the Earth's equator, the **celestial equator** at 0° declination. Right ascension is similar to longitude with the exception that astronomers measure only in one direction, not east and west as does longitude. Right ascension can range from 0 to 24 hours and is usually expressed in hours, minutes, and seconds. Right ascension is the distance measured eastward along the celestial equator from the Sun's position of the first day of spring. A reference is generally required to locate objects using the right ascension/declination system.

Astronomical distances can be huge! The average distance between Earth and Sun, 93 million miles, is called an **astronomical unit** or **AU**. Distances in AU are not always convenient when measuring astronomical distances. A larger distance unit is the **light-year**. A light-year is the distance light travels in one year with a speed of 186,000 miles per second. This distance is 6 trillion miles or 6×10^{12} miles. An even larger distance unit is the **parsec** (parallax-second). A parsec is 3.26 light-years.

Geology

Like other planets in the solar system, Earth began as a cloud of gas and dust spinning around the Sun. In areas where the material was more concentrated, gravity caused additional material to clump together. Greater pressure produced an increase in temperature until the material was a ball of molten rock. As Earth started to cool, rocks solidified. Earth has a diameter of about 12,800 km or 7,930 miles.

The Earth has a layered structure consisting of three major parts with four major constituents each differing in chemical composition. The **crust** is a thin layer that may be classified as oceanic crust with a thickness of about 5 km or continental crust with a thickness of 20-40 km but can be up to 70 km thick under mountains. Oceanic crust is primarily made up of basalt. Continental crust is mostly composed of granite. Granite is less dense than basalt. The **mantle** is mostly solid rock but does contain pools of hot, molten rock called **magma**. The mantle makes up about 80% of Earth's volume. The uppermost mantle, like the crust, is relatively cool and rigid. The mantle is about 2,885 km thick. The crust and uppermost mantle make up the **lithosphere**. The lithosphere averages about 100 km in thickness and acts like a single unit although the crust and mantle have different chemical compositions. Below the lithosphere but still a part of the upper mantle is the **asthenosphere**. Temperature and pressure increase with depth below the Earth's surface. Pressure and temperature conditions exist such that between one and about five percent of the asthenosphere is molten. The lithosphere is able to move independently of the asthenosphere. The lithosphere "floats" or slides on the soft plastic rock of the asthenosphere. Increasing temperature with depth causes the transition from hard rock of the lithosphere to plastic rock of the asthenosphere. At the base of the asthenosphere, the strength of the mantle increases due to increasing pressure. The mantle is capable of slow movement over long periods of time. Some have suggested that the lowermost mantle might also be partially molten. The **core** is made up of a **solid inner core** and a **liquid outer core**. The core is thought to be composed of mostly iron and nickel. The outer core is about 2270 km thick. It is molten. The inner core is solid with a radius of about 1220 km or a diameter of about 2500 km.

A **rock** is a naturally formed solid made up of a mixture of minerals (see More Physical Science Reading for more information on minerals). The **rock cycle** illustrates the origin and connections among the three basic rock classes. **Igneous rocks** form when molten rock, either as magma underground or as lava at the surface, cools and solidifies (crystallizes). **Sedimentary rocks** form when preexisting rocks are weathered, transported, deposited, and lithified, i. e., cemented and/or compacted. **Metamorphic rocks** form when heat, pressure, and/or chemically active fluids change preexisting rocks. To determine whether a rock is sedimentary, metamorphic, or igneous in origin it is important to observe a fresh surface. Also, it is helpful to know if the rock is local in origin (i. e., native to your geographic area). The three major rock classes can be distinguished from one another based on various features. Helpful generalizations to distinguish among the three rock classes are listed in the Figure 5-1 Features to Identify Rocks that follows. Remember that exceptions to the generalizations do exist and you should observe a fresh surface.

Textures and composition are used to identify specific rock types. Texture refers to the size, shape, and boundary relations between adjoining minerals in a rock.

Figure 5-1   Features to Identify Rocks

Feature	Igneous Rock	Metamorphic Rock	Sedimentary Rock
"general" appearance – fresh surface	shiny with multiple colors except obsidian, basalt, and gabbro which can be one solid color	multiple colors to one solid color (see "close" appearance below)	one solid color that ranges from tans to black/grays to light pinks to light greens
a "closer" look – fresh surface	rarely has layers, bedding, or foliation and "never" has fossils	may be foliated with minerals roughly oriented or lined up; nonfoliated may have a "sugary" or sparkly appearance; fossils are rare	has layers if the sample is big enough and fossils are possible
individual mineral grains	yes, unless the rock has a glassy appearance	minerals have been metamorphosed, i. e., affected by pressure, heat, or chemically active fluids	not usually except for sandstone, conglomerate, and breccia
use	expensive building faces, tombstones, railroad aggregate	expensive building faces, tombstones, railroad aggregate	"cheap" building faces, road aggregate, cement

Composition refers to the minerals or elements that make up the rock. A summary of composition and texture information follows in Figure 5-2 Rock Compositions and Textures.

Soil is a term defined differently by various professionals. Soil scientists define soil as the upper layers of regolith (layer of rock and mineral fragments on the Earth's land surface) that support plant growth. Factors that affect the formation of soil include: climate (most influential), parent material, time, plants and animals, and slope. A soil profile is made of layers known as horizons. Topsoil includes an O horizon made up of decayed organic matter and the A horizon made up of mineral matter mixed with some humus (decomposed organic matter). Below the A horizon is a layer consisting of mainly light colored mineral particles. Water has carried finer grained particles away from this layer. Below this layer is a layer made up of mostly clay particles transported by water from above. Next is a layer of partially altered parent material that sits on top of unaltered parent material.

Fossils are traces, imprints, or remains of organisms preserved from the geologic past. Fossilization is a relatively rare event. Typically decomposers eliminate all traces of dead organisms before the fossilization process can begin. Fossilization is a process that is favored by rapid burial and when organisms possess hard parts because each hinders decomposition. Index fossils indicate the ages of rock and are useful when they are widespread geographically and limited to a short span of geologic time.

Figure 5-2   Rock Compositions and Textures

Rock Classification	Composition	Textures
Igneous	8 elements make up more than 90% of most igneous rocks – oxygen, silicon, aluminum, iron, calcium, sodium, potassium, and magnesium most of these elements form feldspars, olivines, pyroxenes, amphiboles, quartz, and mica	*phaneritic*-individual crystals big enough to be viewed with the eye *aphanitic*-individual crystals so small they cannot be seen without a microscope *glassy*-similar to glass; may be massive or threadlike *porphyritic-phaneritic*-2 sizes of crystals; both can be seen with the eye; smaller crystals surround larger crystal called phenocrysts *porphyritic-aphanitic*-phenocrysts are set in aphanitic matrix *fragmental*-broken fragments of ejected igneous material; fine dust to large blocks
Sedimentary	most sedimentary rocks are made up of materials abundant in other rocks and are stable under surface pressure and temperature conditions common are: quartz, calcite, clay minerals, and rock fragments	*clastic*-fragments of rock debris classified according to grain size *crystalline*-chemical precipitates that are usually made up of one dominant mineral; interlocking crystals form a dense rock *oolitic*-calcium carbonate precipitated on a sea floor in the shape of small spheres *skeletal*-skeletal fragments of organisms concentrated near a beach or reef
Metamorphic	due to their origin, metamorphic rocks will have a composition close to that of their parent rock whether it be igneous, metamorphic, or sedimentary	*foliation*-is an element of "flatness" found in some metamorphic rocks; it can be: slaty is expressed by closely spaced fractures; schistosity is expressed by parallel arrangement of slightly flattened minerals; gneissic is expressed by alternating layers of different minerals *nonfoliated*-typically composed of only one mineral; appears structureless because stresses during metamorphism may produce hard to see features

The **Geologic Time Scale** is a chronological arrangement of geologic time subdivided into blocks of time. Eons, eras, periods, and epochs represent blocks of time from largest to smaller. The time scale was developed using relative dating principles. The principles used include:

1. Law of Superposition – In undisturbed sedimentary rocks, each bed or layer is older than the bed above it and younger than the bed below it. The age of rocks becomes progressively younger from bottom to top.
2. Principle of Original Horizontality – Layers of sediment are generally deposited in a horizontal position. Therefore, most sedimentary rocks started out with layering close to being horizontal.
3. Principle of Cross-cutting Relationships – A rock or feature must exist before anything like a fault or magma intrusion can intrude into it or cut across it.
4. Principle of Faunal Succession – Fossil organisms succeed one another in recognizable sequences. Sedimentary rocks of different ages contain different fossils and rocks of the same age contain similar fossils. Therefore, any geologic time period can be recognized by its fossil content.

The Theory of Plate Tectonics proposes that the lithosphere is divided into several plates that slowly move relative to one another by "floating" or sliding over the asthenosphere. The plates interact in various ways with other plates to produce earthquakes, volcanoes, mountains, and crust. This theory offers a single model to explain how and why continents move, the sea floor spreads, mountains rise, earthquakes shake, and volcanoes erupt.

Science usually moves forward by little discoveries each requiring months or years of hard work. Occasionally, major scientific revolutions are the result of scientists gathering little advances into a new idea or a new way of looking at old ideas. Modern earth scientists are lucky to be involved at a time when the effects of plate tectonic theory are as important to earth science as Einstein's theory of relativity was to physics.

Plate tectonic theory states that the Earth's outer layer is a 100 km thick shell of rigid rock called the lithosphere which is broken into pieces called **plates**. The plates move horizontally across the Earth's surface by sliding on the asthenosphere.

A **plate boundary** separates one plate from another plate. At the plate boundaries, the plates can: 1) collide into one another, 2) move away from one another, or 3) move horizontally past one another. A plate boundary is a geologically active place where earthquakes, volcanoes, mountain building, and other geological activities and features concentrate.

Shapes of the continents caused Alfred Wegener (1880-1930) in 1915 to propose that the continents at one time were joined. This idea had been suggested by others. Evidences that support the idea of continental drift include: 1) fit of continental coastlines, 2) fossil evidence, 3) paleoclimatology evidence, and 4) geologic evidence. Scientists were skeptical because Wegener could not suggest a plausible mechanism to drive the motion. Today, unequal distribution of heat within the Earth is the underlying driving force for plate movement. However, no one proposed driving mechanism can explain all aspects of plate motion. Later, additional evidences were added that would eventually contribute to a more encompassing theory. Reversals of the Earth's magnetic field and sea-floor spreading data were important to synthesizing ideas of continental drift and sea-floor spreading into plate tectonic theory. In 1962, Harry Hess (1906-1969) is credited with founding the study of plate tectonics. Hess brought together the theory of continental drift, the concept of convection currents, and the nature and behavior of mid-ocean ridges to found the study of plate tectonics.

There are three main types of plate boundaries. **Divergent boundaries** are where plates move apart or diverge resulting in material from the mantle rising to the surface. With plates of oceanic crust, a mid-ocean ridge forms. New seafloor is created in this manner. With plates of continental crust, a rift valley is produced. **Convergent boundaries** are where plates run into one another. If there is a significant density difference between the two plates, the more dense plate is subducted (forced below) the less dense plate. If the plates are the same density, mountains results as continental crust "piles up." If two oceanic plates are involved, island arcs and ocean trenches are produced. If two continental plates are involved, mountains are produced. If one oceanic and one continental plate are involved, mountains and ocean trenches result. Transform or **transform fault boundaries** are where plates grind or slide past one another. Earthquakes are the main geologic event associated with transform boundaries.

An **earthquake** is a vibration of the Earth produced when built up stress results in the rapid release of energy. The energy released radiates in all directions from the earthquake's **focus**, or the location of its origin. A **fault** is a crack where appreciable movement along a fracture has occurred. An **epicenter** is the location on the Earth's surface that lies above the focus of an earthquake. Geologists use differences in arrival times of different types of seismic waves (primary, secondary to determine locations; surface waves are another type of wave) produced by earthquakes to determine earthquake locations.

A **volcano** occurs where hot liquid rock from the Earth's interior moves through the crust and erupts on the surface. At the surface the molten rock is called **lava** but below the surface it is called **magma**. Spectacular volcanic eruptions can occur where tectonic plates collide. Rock pushed far down into the Earth heats up and melts. This melted rock is **viscous**, i. e., it does not flow easily. Viscosity is the resistance to flow. A viscous fluid does not flow as easily as a fluid that is not viscous. Magma that it less viscous, i. e., the magma is "thinner," may result in volcanoes that ooze lava frequently. Very thin magma may produce flood basalts. In contrast, viscous magma can plug up a volcano's vent. Then pressure builds up resulting in an explosive eruption that releases lava, ash, and gases. Volcanoes can differ in their size, shape, and structure.

Mass wasting refers to the downslope movement of Earth material such as soil, rock, or regolith due to gravity. Mass wasting can be spectacular events like landslides or more gradual events such as creep. Tilted mailbox poles, tipped retaining walls, and trees with curved trunks are evidences of mass wasting.

Hydrology
(Note: See the Water section of the Chemistry Reading for additional background information on water.)

More than 97% of the water on Earth resides in the oceans. Less than 3% of the water on Earth is freshwater. Most of Earth's freshwater is stored in glaciers (about 2%

of the total). More freshwater exists as groundwater (about 0.63%) than as surface water in lakes and rivers (about 0.01%).

The **hydrologic cycle** or water cycle is the continuous circulation of water among the oceans, land, atmosphere, and biosphere. The cycle is powered by energy from the Sun. Water evaporates principally from the oceans to the atmosphere where winds transport moisture-laden air. Certain atmospheric conditions are required to cause the moisture to condense into clouds and fall to the ground in various forms of precipitation.

Water that falls onto land may runoff into lakes and streams or may seep into the ground. Water that seeps into the ground may continue downward until it reaches a zone where all open spaces in the sediment and rock are filled with water. This is the **zone of saturation** and water within this zone is known as **groundwater**. The top of this zone is called the **water table**. Above the water table where the rock, sediment, and soil are not saturated is called the **zone of aeration**. An **aquifer** is rock strata or sediment that transmits groundwater freely. Aquifers depend on being porous (having open spaces) and permeable (a reference to the size and "connectedness" of the pore spaces).

Floods result when the discharge of a stream is great enough to exceed the capacity of he channel and overflows its banks. Floods are the most common geologic hazard. Large amounts of precipitation in a relative short period of time or rapid melting of snow cause most floods.

As a generalization, every 12 hours or so, the ocean rises in a high tide, then falls later in a low tide. **Tides** are caused by the Moon's gravity (the Sun's gravity has a lesser effect) pulling on the water in the oceans and creating a bulge or "extra amount of water" on each side of Earth. The Moon pulls the ocean water towards it. The bulges represent high tide. Areas where water was taken or removed (to create the bulge) represent locations of low tide. High tide occurs on the side of Earth facing toward the Moon and low tide on the opposite side. When the Moon and Sun line up their gravitational forces combine to produce an extra high tide and an extra low tide. These are called spring tides. When the Sun and Moon are at right angles, their gravitational forces pull against one another producing tides that are not as high as normal high tide nor as low as normal low tide. These are called neap tides.

Meteorology

Weather is the state of the atmosphere at a given time. **Climate** describes weather conditions over time for a place or region. Earth's **atmosphere** is the gaseous portion of the planet and is held by gravity. The atmosphere thins rapidly with altitude. Earth's atmosphere is 78% nitrogen and 21% oxygen. All other gases, including, argon, carbon dioxide, neon, helium, water vapor, and others account for the remaining one percent. The atmosphere can be divided into layers. The bottom layer is the **troposphere**. The troposphere goes up about 12 km (of 80 to 300 km) but contains about three-fourths of the atmosphere's gases. Most weather occurs in the troposphere. Energy from the Sun warms the troposphere but it gets most of its heat energy indirectly from the ground. An important component of the atmosphere is **ozone**. Ozone is a form of oxygen made up of three oxygen atoms compared to the oxygen we breathe which is made up of two oxygen atoms. Ozone is concentrated in the stratosphere, a layer of the atmosphere between 10 and 50 km above the surface. Ozone absorbs ultraviolet (UV) light. Ultraviolet light in small amounts produces a "suntan" in people. Large amounts of UV can cause skin cancer, cataracts, and inhibit the growth of many plants. Studies have indicated that compounds containing chlorine and fluorine, called chlorofluorocarbons or CFCs rise into the atmosphere and destroy ozone. Some nations have limited their use of CFCs in an attempt to reduce ozone depletion.

For many, temperature is the most important component of weather. **Temperature** is a measure of the degree of hotness or coldness of a substance. Some define temperature as a measure of the average kinetic energy of individual atoms or molecules in a substance. Temperature can be measured in degrees Fahrenheit, degrees Celsius, or Kelvin. Absolute zero is 0 K or –273.15 °C or –459.67 °F. Under certain defined conditions, water boils at 373.15 K or 100 °C or 212 °F.

Heat can be transferred in three ways: **conduction**, **convection**, or **radiation**. Conduction is transfer of heat energy when one molecule, atom, or particle of greater kinetic energy collides with a particle with less kinetic energy. The heat energy flows from the higher temperature to the lower temperature. Convection is the transfer of heat energy in fluids (gases or liquids) when a warm fluid rises because it is less dense and cooler, more dense fluid sinks. **Differential heating** produces density differences. As long as the fluid is heated unevenly, **convective circulation** or **convection currents** are produced. Radiation is the transfer of heat energy through space by electromagnetic waves. **Land** and **sea breezes** on a daily basis and **monsoons** on a seasonal basis are related to differential heating and pressure differences.

An **air mass** is a large body of air that has similar moisture and temperature characteristics at the same altitudes. Air masses are classified based on their region of origin. Moisture content is designated with a **c** for continental or relatively dry or an **m** for maritime or marine or relatively wet. **P** for polar, designates relatively cool. **T** for tropical, designates relatively warm. The various combinations include: cP, cT, mP, and mT. When an air mass moves out of its area of origin, it will take its moisture and temperature characteristics with it.

Fronts are boundaries that separate different air masses. A **warm front** is shown on a weather map as a line with semicircles extending in the direction that the front is moving. Behind a warm front the air is relatively warmer than the air in front of the warm front. Winds behind a warm front are usually from the south or southwest. The average slope of a warm front is about 1:200. As a warm front approaches, the sequence of clouds generally includes: cirrus, cirrostratus, altostratus, thicker stratus, and then perhaps nimbostratus. Because of their low slopes and relatively slow rate of movement, "bad" weather associated with warm fronts tends to be less severe but longer in duration than bad weather associated with a cold front. A **cold front** is shown on a weather map as a line with triangles extending in the direction that the front is moving. Behind a cold front the air is relatively cooler than the air in front of the cold front. Winds behind a warm front are usually from the north, northwest, or sometimes west. The average slope of a cold front is about twice that of a warm front or about 1:100. As a cold front approaches towering cumulus or cumulonimbus clouds can often be seen at a distance. Because of their relatively high slopes and relatively fast rate of movement, "bad" weather associated with cold fronts tends to be more severe but shorter in duration than bad weather associated with a warm front. An **occluded front** is formed when a faster moving cold front overtakes a warm front. Warm air gets trapped between two colder air masses. Precipitation occurs along both frontal boundaries, combining the narrow band of heavy precipitation associated with a cold front with a wider band of lighter precipitation associated with a warm front. A **stationary front** occurs along the boundary between two stationary air masses. The surface position of the front does not move. The front can remain over an area for several days. Conditions may be similar to those in a warm front. As a result, rain, drizzle, and fog may occur.

Pressure centers are identified on weather maps with "**H**" for **high pressure** or **anticyclones** and "**L**" for low pressure or **cyclones**. Low pressure is associated with rising air and the potential for cloud formation and precipitation exists. High pressure is associated with falling air and this subsidence tends to produce clear skies. Air converges around a low and diverges around a high which is one way of saying that air moves from areas of higher pressure to areas of lower pressure. A **barometer** is used to measure atmospheric pressure. Air must flow toward a low pressure region to replace rising air. Sinking air displaces surface air pushing it out from a high pressure region. Vertical air flow is accompanied by horizontal air flow called **wind**. Winds near the Earth's surface always flow toward a low pressure region from a high pressure region. Wind is caused by pressure differences resulting from the unequal heating of the Earth's atmosphere. **Jet streams** are relatively narrow bands of high speed winds in the upper atmosphere. Jet streams blow from west to east but can move northward or southward. They form when strong temperature differences create big pressure differences at high altitudes.

The National Weather Service uses a combination of heat and humidity called the **heat index** or apparent temperature to measure the dangerous combination of high temperatures and humidity. Other factors such as cloudiness help determine the heat danger. In cold weather, the **wind-chill index** factors in temperature and wind speed to make a determination of how cold it feels to living things.

Clouds are a form of condensation dense enough to be visible to the eyes and made of suspended water droplets or ice crystals. Differential heating of the surface may cause the air over a warmer surface to gain heat and increase in temperature. The warmer air may expand, become less dense and rise. The less dense air moves into areas of lower pressure at higher elevations. The response of the warm, less dense air when it enters areas of lower pressure is to expand. When it expands, the air cools **adiabatically**. An adiabatic temperature change is a change in temperature without a change in energy. If the air cools to the dew point temperature and below, clouds may form. Clouds can be classified according to height and form. The three basic forms include: cirrus, cumulus, and stratus.

Cirrus clouds are high, white, and have a thin, wispy appearance. Cirrus clouds are made of ice crystals. Generally, cirrus clouds are not precipitation makers. When cirrocumulus clouds follow cirrus clouds and the sky coverage increases, stormy weather may be approaching. **Cumulus** clouds generally have a flat base and have a billowy appearance illustrating rising air. **Stratus** clouds are layers or sheets that cover most or all of the sky.

High clouds (bases above 6000 meters or 20,000 feet) include cirrus, cirrostratus, and cirrocumulus clouds. Middle clouds (bases 2000-6000 meters or 6500-20,000 feet) have the prefix "alto" added to their names. Altocumulus and altostratus may be accompanied by light snow or drizzle. Low clouds (bases below 2000 meters or 6500 feet) include stratus, stratocumulus, and nimbostratus. On occasions stratus clouds may produce light precipitation.

Fog is a cloud with its base near or at the ground. The main difference between fog and a cloud is the method and place of formation. Clouds result from differential heating of the Earth's surface that causes less dense air to rise and cool adiabatically. If the air cools below the dew point, a cloud can form. Most fogs are the result of radiation cooling (radiation heat loss by Earth) or the movement of air over a cold surface. The exception is upslope fog. Advection fog is produced when warm, moist air moves over a cold surface. Radiation fog forms on calm, cool, clear nights when the Earth's surface cools rapidly by radiation. Air close to the ground is cooled below its dew point. Cool, more dense air flows into low spots resulting in fog in low lying areas. When relatively warm air is forced to move up a sloping plain or mountain slope the air expands and cools adiabatically. If cooling below the dew point occurs, fog may form.

Relative humidity is the amount of moisture in the air compared to how much moisture the air can hold at a given temperature. It is usually reported as a percent. From this definition, the most obvious way to change the relative humidity is to change the amount of moisture in the air. This is done in homes by using a humidifier to add moisture to the air or a dehumidifier to take moisture out of the air. The less obvious way to change the relative humidity is to change the temperature. If all other variables are kept the same and since warm air can hold more moisture or water vapor than cold air, increasing the temperature has the net effect of decreasing the relative humidity.

Dew point temperature is the temperature at which condensation or precipitation will occur. **Condensation** is a change of state from a gas to a liquid. **Precipitation** is any form in which atmospheric moisture returns to the Earth's surface. For precipitation to occur, three requirements must be met. Moisture, cooling air, and condensation nuclei (surfaces) are all necessary. Air must be cooled to the dew point. If cloud droplets collide and coalesce they may become big enough to fall as mist (0.005 to 0.05 mm diameter), as drizzle (0.1 to 0.5 mm diameter), or as rain (0.5 to 2.0 mm diameter). It takes about one million cloud droplets to combine to form one average rain drop. Sleet (0.5 to 5 mm diameter) forms when raindrops freeze while falling through a layer of subfreezing air. Glaze, or freezing rain, (layers 1 mm to 2 cm thick) is produced when supercooled raindrops freeze on contact with solid objects. Cloud droplets do not always freeze at 0 °C as expected. Small droplets can remain liquid in a cloud even when the temperature is –40 °C (-40 °F). Such water is supercooled. Supercooled water will easily freeze when sufficiently agitated. Snow (1mm to 2 cm diameter) is produced in supercooled clouds where water vapor is deposited as ice crystals that remain frozen as they fall to the ground. Hail (5mm to 10 cm or larger) is produced in large convective, cumulonimbus clouds where frozen ice particles and supercooled water coexists.

Thunderstorms are familiar severe weather events. Typical thunderstorms occur in three stages. The cumulus stage involves air rising, cooling, and condensation creating a cumulus cloud. In the mature stage, latent heat, i. e., the heat released or absorbed by a substance during a change in state, energizes the storm forming heavy rain and violent winds. In the dissipating stage, the cloud cools, convection weakens, and the storm wanes. **Lightning** occurs when electricity travels between areas of opposite electrical charge. This could be between clouds or from a cloud to the ground. Scientists have not determined how the negative and positive charges build in different parts of a cloud.

Thunder is produced from lightning. Lightning heats the air to more than 43,000 °F causing the air to expand. When the expanding air cools, it contracts. The quick expansion and contraction of air around lightning causes air molecules to vibrate. The back and forth movement of the air molecules makes sound waves.

A rotating column of rising air, a mesocyclone, is the key to a supercell's longevity and power. Supercell thunderstorms can last for hours compared to an ordinary thunderstorm that might last for an hour. The mesocyclone organizes warm, humid air near the ground and cool, dry air from above. In addition, the mesocyclone supplies the spin that strong tornadoes need. Cloud and rain drops evaporate into the cool, dry air cooling the air even further and making it more dense. The spinning mesocyclone catches some of the descending air and pushes it in the direction of the storm's movement. This creates a mini cold front that pushes up warm, humid air. This feeds into the mesocyclone. Water vapor in the rising air supplies moisture needed for cloud droplets, rain, and hail. Latent heat released by condensation and cooling of dry, high altitude air by evaporation are energy sources. Rain and hail fall in the downdrafts. **Tornadoes** are likely to form where cool, downdraft air spins into the mesocyclone. Tornadoes are the most violent of all storms. The Fujita intensity scale is used to rate tornadoes. The scale goes from F0 light damage with winds less than 72 miles per hour to F5 with winds greater than 260 miles per hour.

A **hurricane** is a tropical cyclonic storm with winds greater than 74 miles per hour. A tropical cyclone may be called a typhoon in the western Pacific and a cyclone in the Indian Ocean. Hurricanes are larger and longer in duration than tornadoes. Hurricanes form over warm oceans where warm moist air contributes to their formation. A local disturbance intersecting a low can lead to rising air with latent heat warming the air further adding energy to the storm.

Most of the visible portion of sunlight is transmitted by the lower atmosphere and absorbed by the Earth. The warmed Earth emits **infrared radiation** which is **selectively absorbed by atmospheric gases**, **primarily water vapor** and **carbon dioxide**. The absorption of terrestrial radiation adds to the energy content of the atmosphere. This process is called the **greenhouse effect** named for a similar effect that occurs in greenhouses. The absorption and transmission properties of glass are similar to those of atmospheric gases. The glass of a greenhouse allows sunlight to enter the greenhouse and warm the inside of the greenhouse. Glass in a greenhouse will selectively absorb radiation and keep heat trapped inside the greenhouse. A closed car on a cool, but sunny, day illustrates the warming effect of sunlight passing through glass.

Some scientists have expressed concern about human activity that increases the amount of carbon dioxide in the atmosphere. They feel that increased amounts of carbon dioxide in the atmosphere have increased the amount of reradiated energy absorbed and has contributed to **global warming**.

El Nino is the name given to periodic warming of the ocean in the central and eastern Pacific. El Nino affects regional rainfall patterns, modifies global atmospheric circulation patterns, and alters weather on a world-wide scale.

Seasons are caused mainly by the angle that energy from the Sun strikes the Earth. The amount of energy output from the Sun does not change but the angle at which the energy strikes the Earth affects the amount of energy dispersed per unit of area. A lesser factor is the length of daylight. The solar energy is more concentrated, i. e., more energy per unit of area, when the Sun is high in the sky. Also, when the Sun's energy strikes the Earth at close to a 90° angle, the energy must pass through less of the Earth's atmosphere.

Scientists

Aristotle lived from 384-322 B. C. He studied at Plato's Academy, served as a tutor to the future Alexander the Great, and later set up his own school called the Lyceum. His main contribution to science was his emphasis on observation and classification. Aristotle believed a spherical Earth was at the center of the universe and the rest of the universe consisted of a series of spheres arranged around Earth. The sphere nearest to the Earth was the atmosphere and farther out were spheres carrying planets and stars. Around 150 A. D., Claudius Ptolemy (90-170 A. D.) described a system of the universe based on the work of Aristotle. His geocentric theory described a stationary Earth with the Moon, Sun, and planets moving around the Earth in circular paths called orbits. This view of the universe was accepted for nearly 1400 years. Nicolaus Copernicus lived from 1473-1543. His heliocentric theory of the universe had the Sun in the middle with planets moving around the Sun in circular orbits. Tycho Brahe (1546-1601) proposed a system that was a compromise: he stated the planets orbited the Sun and the Sun orbited the Earth. Johannes Kepler (1571-1630) served as Brahe's assistant. Kepler used Brahe's astronomical data to conclude planets moved around the Sun in elliptical orbits. Also, he determined the speed of the planets varied according to their distance from the Sun and that the Sun influenced the movement of the planets. Galileo Galilei lived from 1564-1642. He made observations using telescopes, e. g., mountains on the Moon, sunspots, and Jupiter possessed moons, that played a major role in convincing others that the Copernican view of the Earth revolving around the Sun was correct. His views caused him to be brought before a church court in Rome called the Inquisition. Under the threat of torture, Galileo did publicly deny his claim but is claimed to have said, "And yet it moves!" as he left the court because he believed his theory. Isaac Newton, born the year Galileo died, lived until 1727. Newton used Kepler's laws of planetary motion when he formulated his theories of gravity.

In the late 1700s James Hutton, a Scottish doctor and farmer, published a book that stated geological processes acting today also worked in the past. The expression, "the present is the key to the past" is frequently used to express this idea. The concept that processes that shape the Earth today are those that shaped the Earth in the past is called **uniformitarianism**. By suggesting slower acting, weaker processes acting over long periods of time, Hutton suggested an age of the Earth much older than that suggested by biblical scholars. During the middle part of the 1800s, English geologist Charles Lyell was more successful in interpreting and widely distributing the idea of uniformitarianism than Hutton was earlier. At least some of Lyell's success can be attributed to the fact that Hutton's *Theory of the Earth* was hard to read and understand. By publishing multiple editions of an easier to read and updated *Principles of Geology*, Lyell is given much credit for advancing modern geology.

In a book published in 1915, Alfred Wegener cited evidence for the idea of continental drift. By supporting his claims with evidence and publishing them, Wegener had more impact than did others who earlier suggested the idea of continental drift. A German meteorologist and geophysicist, Wegener suggested a supercontinent broke into smaller continents that have moved to their current locations. A major criticism of Wegener's idea of continental drift was the lack of an explanation for a way to have continents move across the Earth.

Science Careers

Earth and space science careers represent a wide range of opportunities that focus on Earth-related issues from resource management to environmental protection. Earth and space scientists have made many important contributions to society. Career opportunities exist in education, industry, and government. Many geoscientists work in the petroleum industry, mining industry, for environmental companies, or for government agencies. Because earth and space science is made up of a number of disciplines, career opportunities cover a wide range of possibilities such as: astronomer, geologist, hydrologist, meteorologist, oceanographer, and soil scientist. A college degree is required and some jobs require advanced degrees. For more information, contact the American Geological Institute or the American Meteorological Association.

Earth and Space Science Case Study – Plate Tectonics

Plate tectonics is a theory that explains the origin of most of Earth's surface features and many geologic events based on the types of plates and their interactions. Plate tectonics does for geology what evolution does for biology. The acceptance of this theory depended not only on the evidence that supports it but also on its power to explain geologic features and events. It took an accumulation of evidence over a period of time for many scientists to accept the idea of moving continents.

The idea of continental drift was suggested by matching the eastern coast of South America with the western coast of Africa. Alfred Wegener reintroduced the idea early in the 20th century with additional evidence provided by rock types and fossil indicators. The theory of plate tectonics gained wide acceptance in the 1960s when even more evidence supported it. Such evidence included: demonstrating that solid rock could flow under certain conditions, less than expected amounts of sediment on the ocean floors, magnetic evidence, earthquake locations, and seafloor spreading.

Earth and Space Science Case Study – Extending Time

Until the 19th century, most people believed the Earth to be a few thousand years old. Scientists in the 1800s proposed the age of the Earth to be many millions of years old based on indirect evidence from rock formations and fossils. In publishing *Principles of Geology*, Englishman Charles Lyell advanced the notion that Earth might be much older than thousands of years. Lyell used lots of observations and logic to draw inferences from his data. Charles Darwin, in formulating and presenting his theory of evolution, used Lyell's method of supporting his argument with lots of evidence. Currently, most scientists accept a figure of close to 4.6 billion years for the age of the Earth. Scientists now have methods to determine absolute dates where earlier scientists used relative dating methods.

air mass (cP, cT, mP, mT)

altitude

anticyclone

aquifer

asteroid

asthenosphere

astronomical unit (AU)

astronomy

atmosphere

azimuth

barometer

Big Bang Theory

black hole

carbon

carbon dioxide

circulation

climate

cloud (cirrus, cumulus, stratus, formation)

cold front

comet

condensation

conduction

constellation

convection

convection current

core (inner, outer)

crater

crust (continental, oceanic)

cyclone

day

declination

density

dew point

differential heating

dust

$E=mc^2$

Earth

earthquake

eclipse (lunar, solar)

elements

ellipse

El Nino

energy

epicenter

fault

fission

flood

focus

fog

fossil

front

fusion

galaxy

gas

geologic map

geology

global warming

gravity

greenhouse effect

greenhouse gases

heat index

helium

Hertzsprung-Russell (H-R) diagram

high pressure

humidity

hurricane

hydrogen

hydrologic cycle

hydrology

illumination

insolation

jet stream

Jupiter

land and sea breezes

lightning

light-year

lithosphere

low pressure

magnitude (absolute, apparent)

mantle

Mars

Mercury

meteor/meteorite/meteoroid

meteorology

Milky Way

mineral

moon (phases, location, time)

NASA

NASA Strategic Enterprises

Neptune

night

occluded front

oceanography

orbit

ozone

ozone depletion

par sec

planet

plate boundaries

plate tectonics

Pluto

precipitation (types, conditions, requirements)

pressure

radiation

red giant

relative humidity

revolution

right ascension

rock cycle

rotation

Saturn

scale model

scattering

seasons

slope

solar system

speed of light

star (formation, existence, destruction)

supernova

system

telescope

temperature

thermometer

thunder

thunderstorm

tides   topographic map   tornado
troposphere   Uranus   Venus
warm front   water cycle   water vapor
wave (P, S, surface) weather   weather map
white dwarf   wind   wind-chill

Part 6   Physics

Name(s)   _____

_____

_____

_____

86   Work and Power

Exploration/Awareness:

In this activity you will investigate the concepts of work and power.

1. Record your name in the table that follows.
2. Determine the vertical height of the stairs from the lowest floor to the highest floor designated by your instructor. Measure this height in meters and feet or measure in one unit and convert to the other.
3. In seconds, time how long it takes for you to travel from the designated lowest floor to the designated highest floor. **Be careful!** Complete this step of the procedure safely. Record the time in the table.
4. In pounds, record your weight or the force in the table.
5. Convert pounds to kilograms using 1 kg = 2.2 pounds. Record your calculation in the table. Then convert kilograms to newtons by multiplying kilograms by 9.8 m/s^2. Record your calculation.
6. Calculate the work using W = F x d in both SI and USCS or English units. Record each in the table.
7. Calculate power in SI, using P = W/t, and in USCS or English units converting to horsepower using P = W/t x 1hp/550 foot pounds/second.

Table 6-1   Work and Power

1. Name	
SI distance 2. d or distance, stair height in meters (m)	
USCS distance 3. d or distance, stair height in feet (ft)	
SI and USCS time 4. t or time, in seconds (s)	
USCS force 5. F or force (weight), in pounds (lbs)	
SI mass 6. Mass, in kg (mass=lbs x 1kg/2.2 lbs)	

SI force
7. F or force, in Newtons (N)
     (N=kg x 9.8 m/s^2)
     (note: F=ma)	

USCS work 8. W or work, in foot pounds (ft lbs) (W=F x d)	
SI work 9. W or work, in newton meters (Nm) (W=F x d)	
SI power 10. P or power, in watts [P = W/t or (N x m)/t)]	

USCS power
11. P or power, in horsepower (hp)
     [P = (ft x lbs)/s x 1hp/(550 ft lbs/s)] or
     HP = (ft lbs)/s x 1HP/(550 ft lbs/s)	

Concept Development:
8. Work depends on what factors?

9. Power depends on what factors?

10. Do lighter or heavier students do more work when climbing the stairs?

11. How does the speed at which the stairs were climbed affect the power?

12. Compare your results with other students. Are the results what you expected? Why or why not?

Application:
13. Describe a set of conditions that would yield 100 watts.

14. How fast would you have needed to move up the stairs to generate 1 horsepower?

Background Information:
     For additional background, read the **Work** and **Power** sections of the Physics Reading.

87   **Speed, Velocity, Acceleration, Momentum, Potential and Kinetic Energy**

Related National Science Education Standards:
  Teaching Standard A: inquiry-based
    Science Content: systems, order, and organization
                                  evidence, models, and explanation
                                  constancy, change, and measurement
                                  evolution and equilibrium
                                  form and function
    Science as Inquiry: abilities to do scientific inquiry
                                  understanding about scientific inquiry
    Physical Science: properties of objects and materials
                                  position and motion of objects
                                  motions and forces
                                  transfer of energy
    Science and Technology: abilities of technological design
                                  understanding about science and technology
    Science in Personal and
    Social Perspectives: science and technology in society
    History and Nature of
    Science: science as a human endeavor
                                  nature of science
                                  nature of scientific knowledge

Exploration/Awareness:

Variations of the activities described in here can be completed using materials that vary in their technological sophistication. Activities described here will make use of readily available materials. If graphing calculators, computers, sensors, air tracks, . . . etc. are available, it may be desirable for you to use the more sophisticated equipment.

1. Think about two spheres, one a marble and the other a bowling ball. What does it take to cause them to move?

2. Once moving, what does it take to make the marble and bowling ball stop moving?

3. Using spheres of approximately the same diameter but with different masses, allow them to roll down a long incline. Do not roll or push the spheres; just release them and allow gravity to cause them to move. Measure the distance they travel and use a stopwatch to time how long it takes the spheres to cover the distance. Record your

data in the table that follows. Complete more than one trial to determine if you get consistent data.

Table 6-2   Distance, Time, and Average Speed

Sphere	Distance	Time	

4. Once you have determined that you have consistent data from step #3, allow a sphere to roll down the incline but mark the location of the sphere at the end of every second it travels. Be a "problem solver" as to how to best accomplish this.

5. Record your time and distance data from step #4.

Accumulating time	Total distance traveled
0	
1	
2	
3	
4	
5	
6	

(continue as needed)

Concept Development:

6. Inertia is the tendency of an object to keep its present state of motion, i. e., if it is moving it will keep moving and if it is not moving it will not move unless an unbalanced force acts on it. What is different about what it takes to cause a marble to move versus what it takes to cause a bowling ball to move?

7. Speed is the distance traveled per unit of time ($s = d/t$). Label the last column in Table 6-2 as "average speed" and calculate the average speed making sure that you include units in your answer.

8. Velocity is the speed and direction of an object ($v = d/t$ plus direction). If you calculate average velocity for the spheres, would the numbers and units be different from those calculated for average speed? Explain.

9. Once moving, an object has momentum. Momentum can be calculated by taking mass times velocity. How does the mass of an object affect its inertia and its momentum?

10. Acceleration is defined as the rate of change of velocity or as the final velocity minus the initial velocity divided by the final time minus the initial time.
    Symbolically, this is: $a = \dfrac{v_f - v_I}{t_f - t_i}$.
    From step #4, calculate the average acceleration for one of the spheres.

11. For one of the spheres, how did the acceleration change? Or, did it not change?

12. Kinetic energy is energy due to the motion of an object and can be calculated using: $KE = .5mv^2$ where if m is the mass in kg and v is the velocity in meters per second, then KE is expressed in joules. Could the KE of a marble be larger than that of a bowling ball? Explain.

13. Potential energy is stored energy or energy due to position or composition. For an object that has been lifted, PE can be calculated using: $PE = mgh$ where m is the mass in kg, g is the acceleration due to gravity or 9.8 m/s^2, and h is the vertical distance in meters through which the object has been lifted. As with KE, PE is expressed in joules. Calculate the PE of one of the spheres when held at the top of the incline.

Application:
14. Be prepared to define the terms and use the formulas introduced in this activity.

Background Information:
    For additional background, read the **Motion, Energy,** and **Kinetic and Potential Energy** sections of the Physics Reading.

88   **Bicycles: Studying a Compound Machine**

Related National Science Education Standards:
  Teaching Standard A: inquiry-based
    Science Content: systems, order, and organization
                                        evidence, models, and explanation
                                        constancy, change, and measurement
                                        evolution and equilibrium
                                        form and function
    Science as Inquiry: abilities to do scientific inquiry
                                          understanding about scientific inquiry
    Physical Science: properties of objects and materials
                                          position and motion of objects
                                        motions and forces
                                        transfer of energy
    Science and Technology: abilities of technological design
                                          understanding about science and technology
    Science in Personal and
    Social Perspectives: science and technology in society
    History and Nature of
    Science: science as a human endeavor
                                        nature of science
                                        nature of scientific knowledge

Exploration/Awareness:

       Bicycles are common compound machines. Compound machines are combinations of six simple machines: lever, wheel and axle, pulley, inclined plane, screw, and wedge. Simple machines help people do work by increasing force, changing the direction of a force, or changing the speed of a force.

       The rear wheel of a 21-speed bicycle has a set of gearwheels or cogwheels of different sizes. The bicycle chain slips sideways from one gearwheel or cogwheel onto the next smaller or larger gearwheel as the rider changes gears. Smaller gearwheels have fewer teeth than larger gearwheels. For each turn of the front gearwheel attached to the pedals, the more times the rear gearwheel and rear wheel of the bicycle turn around. Using the larger rear gearwheels requires less pedaling effort than using smaller rear gearwheels. However, when using a larger rear gearwheel, the bicycle and rider do not move very far for each turn of the pedals. Experienced bicyclists shift gears to maintain a steady pedaling rate.

Mechanical advantage (MA) can be calculated in two ways:

$$MA = \frac{\text{effort distance}}{\text{resistance distance}} \quad \text{or} \quad MA = D_e/D_r$$

$$MA = \frac{\text{resistance force}}{\text{effort force}} \quad \text{or} \quad MA = F_r/F_e$$

For a bicycle, the circumference of the front sprocket is the effort distance and the circumference of the rear sprocket is the resistance distance.

A student observed a 21-speed bicycle and noted that its various parts incorporate a number of simple machines. In addition, the student made measurements that are recorded in Table 6-3 Dimensions of a 21-Speed Bicycle.

Table 6-3   Dimensions of a 21-Speed Bicycle

Sprocket	Sprocket Diameter	Number of Teeth
largest front	16.5 cm	42
medium front	13.0 cm	34
smallest front	9.0 cm	24
largest rear	11.6 cm	28
second largest rear	9.6 cm	24
third largest rear	9.0 cm	21
medium rear	8.2 cm	18
third smallest rear	6.5 cm	15
second smallest rear	5.5 cm	13
smallest rear	4.6 cm	12

Concept Development:
1. Identify the simple machines represented by the parts of a bicycle that follow.
   a. main road wheels

   b. gear shifter

   c. brakes

   d. pedals and large front sprocket or gearwheel (or cogwheel)

   e. small rear sprocket or gearwheel (or cogwheel) and rear main road wheel

f. chain going around sprockets or gearwheels or cogwheels

2. What do the three front sprockets or gearwheels (or cogwheels) functioning with the seven rear sprockets or gearwheels (or cogwheels) produce?

3. If the largest front sprocket is working in combination with the smallest rear sprocket, is it relatively hard to pedal or relatively easy to pedal?

4. Does the combination of sprockets used in #3 result in a relatively high or low rate of speed?

5. Calculate the mechanical advantage when using the largest front sprocket in combination with the smallest rear sprocket.

6. If the smallest front sprocket is working in combination with the largest rear sprocket, is it relatively hard to pedal or relatively easy to pedal?

7. Does the combination of sprockets used in #6 result in a relatively high or low rate of speed?

8. Calculate the mechanical advantage when using the smallest front sprocket in combination with the largest rear sprocket.

Application:
9. The gear ratio can be calculated by dividing the number of front sprocket teeth by the number of rear sprocket teeth. If the chain stays on the same front sprocket, how does the gear ratio change as the chain moves to smaller sprockets in the rear?

10. What combination of sprockets (or gearwheels or cogwheels) results in the largest mechanical advantage?

11. Under what circumstances might the smallest mechanical advantage be of benefit to a bicycle rider?

12. Is there a relationship between gear ratio and mechanical advantage? Provide evidence to support your conclusion.

Background Information:
For additional background, read the **Machines** section of the Physics Reading.

89   **Motion Questions and Problems**

1. In what ways are scalar and vector quantities similar? How are they different? Give an example of each.

2. A swimmer swims in a straight line down and back in a 100m length swimming pool. What are the swimmer's distance and displacement:

   a. on returning to the starting point?

   b. when at the opposite end of the pool?

3. Use the Pythagorean theorem to determine how far a person travels and their displacement from the starting point if the person drives 30 km south and then 40 km west.

4. A sprinter runs a 100 m in 10 seconds. What is the average speed?

5. If an airplane travels west at 450 km in 1.5 hours, what is its average speed? What is its average velocity?

6. During takeoff, a type of airplane achieves a velocity of 120 km/h in 5.0 s. What is its average acceleration (assume 0 km/h as the initial velocity)?

7. In the space below and using axes labeled "distance" and "time," sketch a graph that shows constant speed. On a second graph, using axes labeled "velocity" and "time" sketch a graph that shows constant acceleration.

90   **Relative Motion Questions and Problems**

1. Using scale diagrams, sketch two objects moving at 5 mph with respect to (wrt) one another. Each object must be moving wrt Earth.

2. Using scale diagrams, sketch two objects moving at 7 mph wrt one another with only one of the objects moving wrt Earth.

3. Using scale diagrams, sketch two objects moving at 6 mph wrt one another, with their movement in opposite directions, and with both of the objects moving wrt Earth.

4. Describe how two objects can be moving at the same speed but with different velocities.

91   Light a Bulb

Related National Science Education Standards:	
Teaching Standard A: inquiry-based	
Science Content:	systems, order, and organization
	evidence, models, and explanation
	constancy, change, and measurement
	form and function
Science as Inquiry:	abilities to do scientific inquiry
	understanding about scientific inquiry
Physical Science:	properties of objects and materials
	light, heat, electricity, and magnetism
	transfer of energy
	conservation of energy . . .
	interactions of energy and matter
Science and Technology:	abilities of technological design
	understanding about science and technology
History and Nature of Science:	nature of science

Exploration/Awareness:

In this activity you will use a single wire, bulb, and battery to determine configurations that will light the bulb and configurations that will not light the bulb. **Safety alert: Some configurations that you may attempt may cause the objects used to become warm or hot! Stop! Do not continue with these configurations! Let go! Hold objects by their plastic or glass parts.**

1. Before testing various configurations, make a detailed sketch of the battery and make a detailed sketch of the bulb showing as much detail as possible.

battery   bulb

2. Now, using a single battery, a single bulb, and a single wire, try to light the bulb. Make sketches of arrangements that do not light the bulb and sketches of arrangements that do light the bulb. Once you find an arrangement of the three objects that results in lighting the bulb, find what changes you can make with this arrangement (or these arrangements). You should have at least four sketches that light the bulb and four that do not light the bulb.

do not light the bulb

do light the bulb

Concept Development:

3. In your own words, describe the arrangements that result in lighting the bulb. What do these arrangements have in common? As a group, how do they differ from those arrangements that do not light the bulb?

4. Using very specific language, describe exactly must be done to light a bulb.

Application:

5. The description you have written in #4 above (if correct) is an operational definition for "closed electrical circuit" or, more frequently, "circuit." Why is the word "circuit" appropriate in this context?

6. In what situation would the term "open circuit" be applicable?

7. In what situation would the term "short circuit" be applicable?

Background Information:

    For additional background, read the **Electricity** section of the Physics Reading.

92   **Conductors and Insulators**

Related National Science Education Standards:	
Teaching Standard A: inquiry-based	
Science Content:	systems, order, and organization
	evidence, models, and explanation
	constancy, change, and measurement
	form and function
Science as Inquiry:	abilities to do scientific inquiry
	understanding about scientific inquiry
Physical Science:	properties of objects and materials
	light, heat, electricity, and magnetism
	transfer of energy
	conservation of energy . . .
	interactions of energy and matter
Science and Technology:	abilities of technological design
	understanding about science and technology
History and Nature of Science:	nature of science

Exploration/Awareness:

      In this activity you will place, at some convenient point, a variety of materials in a circuit. **Safety alert: Some configurations that you may attempt may cause the objects used to become warm or hot! Stop! Do not continue with these configurations! Let go! Hold objects by their plastic or glass parts.**

1. Set up a configuration that lights the bulb.
2. One at a time, place various materials in the circuit so that the electricity has the opportunity to pass through the materials. Find at least seven materials that will allow the bulb to stay lighted and seven that will cause the bulb to not light.

<u>Materials that light the bulb</u>   <u>Materials that do not light the bulb</u>

Concept Development:

3. Classify the materials according to their behavior, i. e., allowing the bulb to remain lighted or causing it to go out. What do the materials within each group have in common (other than how they affect the light)?

4. What name is given to the materials that allow the bulb to light, i. e., produce a closed circuit?

5. What name is given to the materials that do not allow the bulb to light, i. e., maintain an open circuit?

Application:

6. Why is gold used as a conductor on satellites but not used frequently as a conductor on Earth?

7. Materials can be classified as conductors or insulators. Do you think there are substances that fall "in between" this classification scheme? Explain.

Background Information:

For additional background information, read the **Electricity** section of the Physics Reading.

93   **Series and Parallel Circuits**

Related National Science Education Standards:
  Teaching Standard A: inquiry-based
    Science Content: systems, order, and organization
                                    evidence, models, and explanation
                                    constancy, change, and measurement
                                    form and function
    Science as Inquiry: abilities to do scientific inquiry
                                    understanding about scientific inquiry
    Physical Science: properties of objects and materials
                                    light, heat, electricity, and magnetism
                                    transfer of energy
                                    conservation of energy . . .
                                    interactions of energy and matter
    Science and Technology: abilities of technological design
                                    understanding about science and technology
    History and Nature of
    Science: nature of science

Exploration/Awareness:

   In this activity you will construct series and parallel circuits and compare the brightness of bulbs in the circuits and between the two types of circuits.

1. Construct a series circuit using two light bulbs, two sockets, wire, and one or more batteries. Draw a circuit diagram using proper symbols. Compare the brightness of the two bulbs. Remove one of the bulbs. What happens?

2. Construct a parallel circuit using two light bulbs, two sockets, wire, and one or more batteries. Draw a circuit diagram using proper symbols. Compare the brightness of the two bulbs. Remove one of the bulbs. What happens?

3. Compare the brightness of the bulbs in the series circuit with the brightness of the bulbs in the parallel circuit.

Concept Development:

4. Explain your results for #1 earlier.

5. Explain your results for #2 earlier.

6. Explain your results for #3 earlier.

Application:

7. Use Ohm's Law to explain your "brightness" results.

8. What are the advantages and disadvantages of the two types of circuits?

9. Provide an example of where each type of circuit is used.

Background Information:

For additional background information, read the **Electricity** section of the Physics Reading.

94   Magnetism

Related National Science Education Standards:
Teaching Standard A: inquiry-based
Science Content: systems, order, and organization
                                evidence, models, and explanation
                                constancy, change, and measurement
                                form and function
Science as Inquiry: abilities to do scientific inquiry
                                understanding about scientific inquiry
Physical Science: properties of objects and materials
                                . . . , electricity, and magnetism
                                properties and changes of properties in matter
                                motions and forces
                                structure of atoms
                                structure and properties of matter
                                motions and forces
                                interactions of energy and matter
Science and Technology: understanding about science and technology
                                distinguish between natural objects and objects made by humans
Science in Personal and
Social Perspectives: personal health
                                risks and benefits
                                science and technology in society
History and Nature of
Science: science as a human endeavor
                                nature of science

Exploration/Awareness:

1. Use the magnets and materials provided to identify seven objects/substances that are attracted to magnets and seven that are not. List them in the space provided.
   CAUTION: Test only the materials provided. Magnets can negatively affect some objects/materials.

Affected by magnets _____   Not affected by magnets _____

2. For those objects/substances attracted to magnets, conduct some tests that determine how distance affects magnetic force and record your observations and conclusions.

3. Use a magnet, iron filings, and other materials (glass sheet or transparency) to create a model that illustrates a magnetic field. Make a sketch of the model that illustrates the magnetic field around the magnet.

4. Use two or more ring magnets and other materials to set up a demonstration where one or more of the magnets will levitate.

5. Rub a paper clip or iron nail with a magnet to see if it can be magnetized.

Concept Development:
6. What types of materials are magnetic or are attracted to magnets?

7. Does a magnet affect copper?

8. What types of materials are not magnetic or are not attracted by magnets?

9. Describe the appearance of the magnetic field produced in step #3. For example, were the iron filings illustrating the magnetic field always the same distance apart? Explain the results.

10. How does distance affect the magnetic force?

11. Explain how one magnet can cause another magnet to levitate.

12. Explain how a magnet can cause another object to become magnetized. Does the other object stay magnetized?

Application:
13. Explain how you could use a magnet to cause another magnet suspended by string to move.

14. A magnet lost its magnetic properties after being heated. Explain how this might have occurred.

15. How might magnetism be used to provide a high-speed form of transportation?

16. Under the guidance of your instructor, conduct an experiment that uses electricity to produce a magnet.

Background Information:
For additional background information, read the **Magnetism** section of the Physics Reading.

95 Cooling

Related National Science Education Standards:
Teaching Standard A: inquiry-based
Science Content: systems, order, and organization
                                    evidence, models, and explanation
                                    constancy, change, and measurement
                                    evolution and equilibrium
                                    form and function
Science as Inquiry: abilities to do scientific inquiry
                                    understanding about scientific inquiry
Physical Science: properties of objects and materials
                                    . . . , heat, . . .
                                    properties and changes of properties in matter
                                    transfer of energy
                                    conservation of energy
                                    interactions of energy and matter
Earth and Space Science: properties of Earth materials
Science and Technology: abilities of technological design
                                    understanding about science and technology
History and Nature of
Science: nature of science

Exploration/Awareness:

In this activity you will use a graphing calculator (GC) and calculator based laboratory (CBL) to investigate cooling.

1. Turn on the calculator.
2. Press the APPS key.
3. Select choice #2, CBL/CBR, by pressing the 2 key.
4. Following on screen directions by pressing any key.
5. Select choice #2, DATA LOGGER, by pressing the 2 key.
6. IF TEMP is flashing, select it by pressing ENTER. If TEMP is not flashing, use the right or left arrow key to highlight TEMP and then select TEMP by pressing ENTER.
7. If not done automatically, use the down arrow key to go to #SAMPLES. Enter 30 (for 30 samples) by pressing 3 then 0 and press ENTER.
8. If not done automatically, use the down arrow key to go to INTRVL (time interval), enter 2, and press ENTER.
9. If not done automatically, use the down arrow key to go to UNITS and select either °C (degrees Celsius) or °F (degrees Fahrenheit) by using the right or left arrow key and then press ENTER.

10. If not done automatically, use the down arrow key to go to PLOT and make a choice by using the right or left arrow key. For us, RealTime will work. Highlight it and then press ENTER.

11. If not done automatically, use the down arrow key to go to DIRECTNS. We want directions to be shown on screen, so use the right or left arrow key to select ON and then press ENTER.

12. Use the down arrow key to proceed to GO. Press ENTER.

13. Follow the on screen directions by connecting the CBL unit to the calculator using a black link cable and then press ENTER.

14. Follow the on screen directions by plugging the temperature probe into the CH 1 port of the CBL unit and then press ENTER.

15. Follow the on screen directions by turning the CBL on and press ENTER.

16. On screen you should read a status message. If OK, proceed by pressing ENTER. If not, you will need to do some trouble shooting such as making sure of connections or get help from your instructor.

17. You should now read a message that says to begin you will need to press ENTER *but do not press ENTER until you follow directions in #18 and #19.*

18. Dip the end of the temperature probe into hot water in a container and hold it there for about 15 seconds.

19. After pulling the probe out of the water, quickly press ENTER to begin data collection. Hold or prop the end of the probe up so air can easily reach the probe.

20. On screen you will see data points marked every two seconds. After data collection is complete, the graph will be plotted automatically.

21. You can press the TRACE key and then use the right and left arrow keys to move across the graph to note particular values.

22. Make sure you have answers for questions 23-26 before you quit. You can quit by pressing ENTER then the 2nd key followed by the QUIT key (it says MODE) and then by pressing 4. Turn the calculator off by pressing the 2nd key and then the OFF key (it says ON). Turn off the CBL by pressing 2nd and then OFF (it says ON/HALT).

Concept Development:

23. The x or independent (manipulated) value represents what variable?

24. The y or dependent (responding) value represents what variable?

25. Record temperatures in Table 6-4 Cooling Data that follows. Use graph paper to construct a graph using the recorded data.

Table 6-4   Cooling Data

Sample Number (x 2 = Time in seconds)	Temperature
0	
2	
4	
6	
8	
10	
12	
14	
16	
18	
20	
22	
24	
26	

28	
30	

26. When does the liquid cool off faster earlier or later in the data collection period?

27. Summarize what this activity illustrates about cooling rates.

Application:
28. You collected data for 60 seconds and the graphing calculator graphed these data. Sketch what you think the graph would look like if additional data were collected. Generally, what would the final temperature be?

29. What would you need to do to collect data every 3 seconds for 1.5 minutes?

30. Briefly describe another activity that would involve collecting time and temperature data.

Background Information:
For additional background, read the **Heat** section of the Physics Reading.

96   Heat and Insulation

Related National Science Education Standards:
  Teaching Standard A: inquiry-based
    Science Content: systems, order, and organization
                                  evidence, models, and explanation
                                  constancy, change, and measurement
                                  evolution and equilibrium
                                  form and function
    Science as Inquiry: abilities to do scientific inquiry
                                  understanding about scientific inquiry
    Physical Science: properties of objects and materials
                                  . . . , heat, . . .
                                  properties and changes of properties in matter
                                  transfer of energy
                                  conservation of energy and increase in disorder
                                  interactions of energy and matter
    Science and
    Technology: abilities of technological design
                                   understanding about science and technology
                                  distinguish between natural objects and objects made by humans
    Science in Personal and
    Social Perspectives: personal health
                                  types of resources
                                  science and technology in science
                                  natural resources
    History and Nature of
    Science: science as a human endeavor
                                  nature of science
                                  nature of scientific knowledge
                                  historical perspectives

Exploration/Awareness:

1. Design and conduct a control variable experiment that determines the insulating properties of materials of your choice. Write up the experiment using the five-step procedure used throughout this course.

Concept Development:

2. What material provided the best insulation?

3. What material provided the worst insulation?

4. List some examples of where materials with good insulating properties are used.

5. What can be done to improve the insulating qualities of a material?

Application:
6. In buildings and homes, insulation is rated according to its "R-value." The R-value stands for resistance to heat flow. What can homeowners do to increase the R-value of the insulation in their homes?

7. How does an igloo made of snow and/or ice keep those inside it warm?

Background Information:
    For additional background, read the **Heat** section of the Physics Reading.

97   Explaining Sunset Colors

Related National Science Education Standards:

Teaching Standard A: inquiry-based	
Science Content:	systems, order, and organization
	evidence, models, and explanation
	constancy, change, and measurement
	form and function
Science as Inquiry:	abilities to do scientific inquiry
	understanding about scientific inquiry
Physical Science:	properties of objects and materials
	light, heat, . . .
	transfer of energy
	structure and properties of matter
	conservation of energy, . . .
	interactions of energy and matter
Earth and Space Science:	properties of Earth materials
	objects in the sky
	changes in Earth and sky
	Earth in the solar system
	energy in the Earth system
History and Nature of Science:	nature of science

Exploration/Awareness:

In this activity you set up a demonstration to illustrate why sunset colors are red or orange. The materials you need include:

2 "colorless," transparent 10 ounce plastic cups – each three-fourths full of water (one will be used for comparison, the other will be used to add whole milk or non-dairy coffee creamer)

flashlight

index card

dropper if you use whole milk

or

5 mL (one teaspoon) of: 5+ mL (one rounded teaspoon) of non-dairy creamer in 375 mL (one and one-half cups) of warm water

1. Place about 6 drops of whole milk into the cup with water and stir or swirl to distribute the milk (you can experiment to determine what number of drops of whole milk works best).
2. Hold the index card behind the cup as you shine the flashlight through the cup with only water. Dimming the room lights may enhance the effect.

3. What color of light shows on the card?

4. Hold the index card behind the cup as you shine the flashlight through the milk and
    water. Dimming the room lights may enhance the effect.
5. What color of light shows on the card?

6. Carefully observe the color of the water and milk solution as you shine the flashlight
    through it. Does it take on the "hint" of a different color (different than when light
    from the flashlight is not shining through it)? If yes, what "hint" of color is it?

Concept Development:
7. What colors make up visible or "white" light?

8. How do you explain the different colors of light that resulted from shining the
    flashlight through water versus the water and milk (or non-dairy creamer) solution?

Application:
9. How might room lights affect the results?

10. Design a control variable experiment that would attempt to determine what amount of
    water and milk (or non-dairy creamer) would produce the best results.

Background Information:
        For additional background information, read the **Light** section of the Physics
Reading.

98   Polaroid Filters

```
Related National Science Education Standards:
    Teaching Standard A:  inquiry-based
    Science Content:              systems, order, and organization
                                  evidence, models, and explanation
                                  constancy, change, and measurement
                                  evolution and equilibrium
                                  form and function
    Science as Inquiry:           abilities to do scientific inquiry
                                  understanding about scientific inquiry
    Physical Science:             properties of objects and materials
                                  light, . . .
                                  interactions of energy and matter
    Science and Technology:       understanding about science and technology
                                  distinguish between natural objects and objects made by humans
    Science in Personal and
    Social Perspectives:          science and technology in society
    History and Nature of
    Science:                      science as a human endeavor
                                  nature of science
                                  nature of scientific knowledge
                                  historical perspectives
```

Exploration/Awareness:

In this activity you will use two polaroid filters to investigate characteristics of light.

1. How does looking through a circular polaroid filter affect the amount of light you can observe?

2. Take a second circular polaroid filter and hold the two filters together. How does the combined effect of the polaroid filters affect the amount of light observed?

3. With both polaroid filters held together, rotate one of the filters so that it makes one complete rotation, i. e., from the starting position one filter is rotated so that moves through 360° and ends up in the original starting position. How does the amount of light change?

4. Rotate the polaroid filters so that a minimum amount of light passes through the filters. About how many degrees does one of the filters have to be rotated to allow the maximum amount of light to pass through the filters? Then from this placement that allows the maximum amount of light, how many degrees does one of the filters have to be rotated to allow the minimum amount of light to pass through the filters?

Concept Development:
5. What might light be like that would explain the results observed in #4?

Application:
6. Other than in activities like this, how are polaroid filters used in the "real world?"

Background Information:
      For additional background, read the **Light** section of the Physics Reading.

99   **Reflection and Refraction**

Related National Science Education Standards:
  Teaching Standard A: inquiry-based
    Science Content: systems, order, and organization
                                     evidence, models, and explanation
                                     constancy, change, and measurement
    Science as Inquiry: abilities to do scientific inquiry
                                     understanding about scientific inquiry
    Physical Science: properties of objects and materials
                                     light, . . .
                                     properties and changes of properties in matter
                                     transfer of energy
                                     structure and properties of matter
                                     interactions of energy and matter
    Science and
    Technology: understanding about science and technology
    Science in Personal and
    Social Perspectives: science and technology in society
    History and Nature of
    Science: science as a human endeavor
                                     nature of science
                                     nature of scientific knowledge

Exploration/Awareness:

This activity will be described using readily accessible materials. Those that have more sophisticated equipment available, e. g., a laser, could use more advanced forms of technology. Follow proper safety procedures!

1. If not already constructed, cut a very narrow slit about half way up a 30 cm by 30 cm piece of cardboard.

2. Stand the cardboard up and hold it in place with a holder, block of wood, or modeling clay.

3. Shine a flashlight through the slit. Place a mirror behind the cardboard at an angle so that you can observe the incoming light and the reflected light. Measure the "angle of the incoming light" with that of the reflected light. Record your measurements.

4. Use a large test tube or a jar like an olive jar and fill it nearly full with water. Place a pencil or straw part way into the test tube or jar so that your finger is immersed in the

water. Describe the appearance of the pencil or straw in the water versus outside the water.

5. Place a coin in an opaque cup. Move your head from a position where you can see the coin to where you just cannot see the coin because the rim of the cup blocks your vision. Hold your head in position while your partner almost fills the cup with water. Can you see the coin again?

Concept Development:
6. How did the "incoming angle of the light" compare to the angle of the reflected light?

7. What must happen to the light to account for your observations in #4 and #5?

Application:
8. Why does a drinking straw placed in a transparent cup holding water appear to be cut in half?

9. List some examples of where the "bouncing back" of light and the "bending" of light are used.

Background Information:
For additional background, read the **Light** section of the Physics Reading.

100   **Distance and Light Intensity**

Related National Science Education Standards:
Teaching Standard A: inquiry-based
Science Content: systems, order, and organization
                                    evidence, models, and explanation
                                    constancy, change, and measurement
                                    form and function
        Science as Inquiry: abilities to do scientific inquiry
                                    understanding about scientific inquiry
        Physical Science: properties of objects and materials
                                    position and motion of objects
                                    light, . . .
                                    properties and changes of properties in matter
                                    transfer of energy
                                    interactions of energy and matter
        Science and
        Technology: abilities of technological design
                                    understanding about science and technology
        Science in Personal and
        Social Perspectives: science and technology in society
        History and Nature of
        Science: science as a human endeavor
                                    nature of science
                                    nature of scientific knowledge

Exploration/Awareness:

In the Cooling activity you developed knowledge and skills associated with using a GC/CBL system. Once again you will use a GC/CBL but instead of measuring temperature you will make selections that allow you to measure light intensity. In this activity you will use a GC/CBL and a light source to determine how distance affects light intensity.

1. Design a control variable experiment that will determine the effect of distance on light intensity.
2. Go through the GC/CBL setup procedures for measuring light intensity.
3. Conduct the experiment and record at least four distances and the light intensity measured at those distances in the space below.

Trial	Distance	Light intensity
1		

Trial	Distance	Light intensity
2		
3		
4		

Concept Development:

4. What are the units used to measure light intensity?

5. Graph your data. Does connecting the data points produce a straight line?

6. How does distance affect light intensity?

Application:

7. What is the relationship between light intensity and distance?

Background Information:

For additional background, read the **Light** section of the Physics Reading.

Name(s)

_____

_____

_____

_____

101   Orientation and Energy per Unit Area

Related National Science Education Standards:
Teaching Standard A: inquiry-based

Science Content:	systems, order, and organization
	evidence, models, and explanation
	constancy, change, and measurement
	form and function
Science as Inquiry:	abilities to do scientific inquiry
	understanding about scientific inquiry
Physical Science:	properties of objects and materials
	position and motion of objects
	light, . . .
	properties and changes of properties in matter
	transfer of energy
	interactions of energy and matter
Science and Technology:	abilities of technological design
	understanding about science and technology
Science in Personal and Social Perspectives:	science and technology in society
History and Nature of Science:	science as a human endeavor
	nature of science
	nature of scientific knowledge

Exploration/Awareness:

In the Cooling activity and in the Distance and Light Intensity activity you developed knowledge and skills associated with using a GC/CBL system. In this activity you will use a GC/CBL and a light source to determine how the orientation of a light sensor affects the energy per unit area.

1. Design a control variable experiment that will determine the effect of angle on energy per unit area, in this case measured as light intensity.
2. Go through the GC/CBL setup procedures for measuring light intensity.
3. Conduct the experiment and record at least four angles and the light intensity measured at the same distances in the space below.

Trial	Orientation	Light intensity
1	light sensor pointed directly at the light source	

Trial	Orientation	Light intensity
2	light sensor pointing less directly at the light source ($\sim 1/6^{th}$ of a rotation)	
3	light sensor pointing even less directly at the light source ($\sim 1/3^{rd}$ of a rotation)	
4	light sensor pointing opposite the light source ($\sim \frac{1}{2}$ of a rotation)	

Concept Development:

4. What are the units used to measure light intensity?

5. Graph your data. Does connecting the data points produce a straight line?

6. Are the results expected? Account for the results.

7. How does orientation of the light sensor affect the energy per unit area or light intensity?

Application:

8. Explain how angle and energy per unit area is related to seasons on Earth.

Background Information:

For additional background, read the **Light** section of the Physics Reading.

102   **Sound**

Related National Science Education Standards:
Teaching Standard A: inquiry-based
Science Content: systems, order, and organization
evidence, models, and explanation
constancy, change, and measurement
form and function
Science as Inquiry: abilities to do scientific inquiry
understanding about scientific inquiry
Physical Science: properties of objects and materials
position and motion of objects
motions and forces
transfer of energy
interactions of energy and matter
Science in Personal and
Social Perspectives: personal health
changes in environments
science and technology in local challenges
risks and benefits
science and technology in society
environmental quality
History and Nature of
Science: science as a human endeavor
nature of science

Exploration/Awareness:

In this activity you will investigate sound. Vibrating objects can produce sounds. Loud sounds represent lots of energy and are loud because the atoms or molecules carrying them vibrate "a lot." The loudness or volume of a sound is the strength of the sound as it reaches the ear. People may perceive the loudness of a sound differently. The decibel scale compares intensities of sounds. Sound intensity depends on the pitch or frequency of the sound wave and the amplitude of the sound wave and is measured in decibels, dB. The faintest sound humans hear is about 10 dB, normal conversation is about 60 dB, dangerously loud music is about 120 dB, and sounds can become physically painful above 130 dB.

1. The bottles with water to be used should already have water in them or be marked to allow you to add water to the appropriate level. Tap on the bottles with the "striker" provided. What is vibrating to produce the sound?

2. Instead of striking the bottles, blow air across the top of the bottle. Now, what is vibrating to produce the sound?

3. As demonstrated by your instructor, how does the "singing glass" produce sound?

Concept Development:

4. How does the amount of water in the bottle affect the pitch of the sound produced by striking the bottle?

5. How does the amount of water in the bottle affect the pitch of the sound produced by blowing air across the top of the bottle?

Application:

6. Why do some towns and cities develop regulations on the maximum decibel levels allowed?

7. "Noise" is sometimes described as sounds produced by irregular vibrations or sound at the wrong place and time. Are these suitable descriptions? Why or why not?

Background Information:

For additional background, read the **Sound** section of the Physics Reading.

Name(s) _____

_____

_____

_____

103   Half-life Simulation

Related National Science Education Standards:	
Teaching Standard A: inquiry-based	
Science Content:	systems, order, and organization
	evidence, models, and explanation
	constancy, change, and measurement
	evolution and equilibrium
Science as Inquiry:	abilities to do scientific inquiry
	understanding about scientific inquiry
Physical Science:	properties of objects and materials
	properties and changes of properties in matter
	structure of atoms
	structure and properties of matter
	interactions of energy and matter
Science in Personal and	
Social Perspectives:	natural hazards
	risks and benefits
History and Nature of	
Science:	science as a human endeavor
	nature of science

Exploration/Awareness:

      This activity is intended to model radioactive decay. Note that nuclear decay is a continuous process but the model used here illustrates a discrete process, i. e., one event occurs, the event is over, and then the next event occurs. In nature, nuclear decay does not stop and start. A half-life is the amount of time it takes for half of a given number of atoms of a radioactive element to decay. The half-life of a radioactive substance is a constant but the half-lives of different radioactive materials can range from fractions of a second to billions of years.

      In this activity, whether or not an atom undergoes nuclear decay will be simulated by an event that has two possible outcomes. Possibilities include flipping pennies where the outcomes are the pennies landing with "heads" up or "tails" up or by shaking and pouring candies like M&M's® on a table where the outcome is "M" side up or "M" side down.

1. Based on the materials provided, decide what event will represent nuclear decay.
2. Count the total number of candies or coins. Write this number in the Half-life Simulation Table as the total number of "atoms" not decayed for the 0 flip/toss/shake.

3. Flip/toss/shake the coins or candies onto a tabletop. Count the number of "atoms" decayed and remove them. Record the total number of "atoms" decayed and the total not decayed for flip/toss/shake #1.
4. Complete another flip/toss/shake using only the undecayed "atoms" from the previous step. Count the number decayed, and it to the prior total, and record in the total decayed column. Remember decayed coins or candies are removed! Also, record the number not decayed.
5. Continue until all "atoms" have decayed. At the end, the total of "atoms" not decayed will be 0 and the total decayed will be the number of "atoms" at the start.

Table 6-5   Half-life Simulation

"Flip" or "toss" or "shake"	Total "atoms" not decayed	Total "atoms" decayed
0		0
1		
2		
3		
4		
5		

Continue as needed

Concept Development:
6. Define half-life.

7. Ideally, for each trial half of the model atoms should have decayed. How closely did the model approximate the ideal?

8. Sketch a half-life graph that shows time or half-lives on the x-axis and amount or number of undecayed atoms on the y-axis.

Application:

9. Use the equation to determine how many model atoms should remain after three half-lives. The "n" stands for the number of half-lives. Show your work in the space provided. How close to the calculated value was the actual value?

amount remaining = $\dfrac{\text{amount in the original sample}}{2^n}$

Background Information:

      For additional background, read the **Nuclear Energy** section of the Physics Reading.

104 Calculating Radiation Dosages

Related National Science Education Standards:
Teaching Standard A: inquiry-based

Science Content:	systems, order, and organization
	evidence, models, and explanation
	constancy, change, and measurement
	evolution and equilibrium
Science as Inquiry:	abilities to do scientific inquiry
	understanding about scientific inquiry
Physical Science:	properties of objects and materials
	properties and changes of properties in matter
	interactions of energy and matter
Science and Technology:	abilities of technological design
	understanding about science and technology
	distinguish between natural objects and objects made by humans
Science in Personal and Social Perspectives:	personal health
	changes in environments
	natural hazards
	risks and benefits
	science and technology in society
	natural and human-induced hazards
History and Nature of Science:	science as a human endeavor
	nature of science
	nature of scientific knowledge

Exploration/Awareness:

We live in a radioactive world. It is a misconception to think that the only sources of nuclear radiation are nuclear fuels like uranium and radioactive sources associated with medicine.

1. Conduct a web search and find a site that allows you to calculate your personal radiation dosage. The site should identify a number of sources of radiation and provide amounts of radiation per year from each source.

2. Record your personal radiation dosage:

personal radiation dosage = _____

Concept Development:

3. What is the average radiation dosage per year? How does your dosage compare to the average?

4. What sources provide relatively high amounts of radiation?

5. What sources provide negligible amounts of radiation?

6. What unit is used to measure radiation?

Application:

7. What can a person do to limit their exposure to nuclear radiation?

Background Information:

     For additional background, read the **Nuclear Energy** section of the Physics Reading.

Name(s) _____

_____

_____

_____

105   **Physics-Related Hobby Report**

Significant numbers of people engage in physics-related activities outside their time at work. They pursue these activities as hobbies out of curiosity, for the enjoyment it brings, and for the benefits it brings to their lives.

What follows is a list of some hobbies or activities where possession of some physics knowledge and/or skills would be useful. Other possibilities exist.

amateur astronomy	biking
car racing	musician/singer
surfing	

Select one of the possibilities from the list or come up with a physics-related hobby or activity of your own (get your instructor's approval on your idea). Write a 1 ½ to 2 page, double-spaced report on the hobby or activity. Address the following in your report:

1. Identify the physics knowledge and skills needed to participate in this hobby or activity. This must be included in the report.
2. Describe the role, if any, that technology plays in this hobby or activity. This must be included in the report.
3. Report the approximate number of people that participate in this hobby or activity. Include this information as available.
4. If references are used, they must be properly cited.

106   Scientists, Nature of Science, and History of Science Report

Physicists have made significant contributions to human history. Select an important physics concept, idea, or theory and identify scientists that contributed to the development of the concept, idea, or theory. Write a 2 to 3 page, double-spaced report on the scientists and topic selected. Address the following in your report:

1. Clearly identify the physics concept, idea, or theory selected. This must be included in the report.
2. Describe the role that scientists played in the development of this concept, idea, or theory. This must be included in the report.
3. Describe how the historical development of the concept, idea, or theory illustrates the nature of science. This must be included in the report.
4. If references are used, they must be properly cited.

107   **Physics Questions and Problems**

1. Calculate the amount of **horsepower** and **watts** produced when a force of 220 pounds is exerted through a distance of 100 m (about 330 feet) in 10 seconds.

2. A speed of 60 mph is about 88 ft/s. If an auto's velocity changes from 88 ft/s to 100 ft/s in 5 seconds while it is traveling northeast, what is the acceleration?

3. If the voltage is 12 volts and the current is 2 amps, what is the resistance?

4. Using conventional symbols, draw a circuit diagram showing 3 bulbs in series.

5. Using conventional symbols, draw a circuit diagram showing 3 bulbs in parallel.

6. If 8 grams of a radioactive substance are left after 6 half-lives (1 half-life = 100 years), how many grams of the radioactive substance were there at the start? Sketch a half-life graph that could be used to illustrate this problem.

7. Compare the amount of kinetic energy and the momentum of a driver in a car with that of a bicycle rider. The driver in the car is traveling at 2.5 times the speed of the bicycle rider and has 15 times the mass.

8. Calculate the mechanical advantage when the resistance force is 120 N and the effort force is 25 N.

(Note: See other Activity-Based Physical Science units for additional reading on selected physical science topics.)

Force

Frequently, a **force** is described as a push or a pull. An unbalanced or net force can cause an object to change its state of motion. This description of a force is a simplification but it does serve as a starting point. An object at rest may be caused to move or an object in motion may change its direction or amount of motion if sufficient force is applied. The source of a force may be muscular, electrical, magnetic, or gravitational. Earlier in this course, weight was identified as the force due to gravity acting on an object. The English or USCS unit for force is the **pound** and the SI unit of force is the **newton** (N).

Work

"Work" is a commonly used term. In science, work has a specific definition that does not always match its everyday usage. In science, **work** is defined as the product of the **force applied** to an object times the **distance** the object moved ($W = Fd$). Force, or F, is the average or constant force acting on a body through a given distance. The d or distance refers to the distance traveled by the body in the direction of the applied force while the force is being applied. SI units for distance include the meter and the SI unit for force is the newton, so the SI unit for work is the newton-meter (N-m). More simply, a newton-meter is called a **joule** (J). One joule (pronounced to rhyme with "cool") is the amount of work or energy needed to maintain a force of one newton through a distance of one meter. Joules are units for work and energy. An English or USCS unit for distance is the foot (or feet) and the unit of force is the pound, so the USCS unit of work is the foot-pound.

Power

In science, **power** is the amount of work done per unit of time, $P=W/t$. The SI unit of power is the **watt**. One watt is equal to one newton-meter (or joule) of work per second. A kilowatt is 1000 watts. The English or USCS unit of power is called **horsepower**. One horsepower is equal to 550 foot-pounds of work per second.

Machines

A machine makes work easier by changing the amount, direction, or speed of an applied force. The number of **simple machines** listed depends upon the source. Some sources list six simple machines: lever, pulley, wheel and axle, inclined plane, screw, and wedge. Other references consider a wedge to be two inclined planes combined and a screw to be an inclined plane that winds around a central point. **Compound machines** are machines that are made up of two or more simple machines.

Mechanical advantage is how many times a machine increases an applied force. In many circumstances people want machines to make work easier by multiplying the force that humans can apply. However, when a gain in the applied force is used to increase the speed of a machine or to change the direction of an applied force, the mechanical advantage may be less than one. **Friction** decreases the mechanical advantage of a machine. Friction opposes motion, causes machine parts to slow down, and causes machine parts to wear. For these reasons, oil or some other lubricant is used to reduce friction and the effects of friction. Friction produces heat and reduces the mechanical advantage of machines.

The **efficiency** of a machine is the ratio of work output to work input. Percent efficiency is calculated by dividing work output by work input and then multiplying by 100% or:

$$\% \text{ efficiency} = \frac{\text{work output}}{\text{work input}} \times 100$$

Reducing friction can increase the efficiency of a machine. Besides using lubricants, another way to reduce friction is to use smooth surfaces. Increased efficiency helps conserve natural resources by requiring less fuel to do a given amount of work.

Motion

A force can cause motion. Balanced forces are forces that are equal in amount but opposite in direction. Balanced forces will not produce motion. Unbalanced forces are situations where there is a net force that can produce motion. All motion is relative, i. e., motion is determined relative to some reference point. If you are sitting absolutely still in a chair, it does not feel like you are moving. However, because you are on a rotating Earth, an observer able to view your position from out in space may detect your motion. **Relative motion** is often described as being motion with respect to (wrt) a reference point. For example, the Earth moves with respect to the Sun. A book that is not moving with respect to the table it lies on is moving relative to the Sun.

An object in motion travels a certain distance in a given amount of time. Speed is a measure of how fast something is moving. **Speed = distance/time** or s=d/t. Any distance unit and time unit can be used. **Instantaneous speed** is the speed that an object has at any instant. Since many moving objects rarely travel at constant speeds, **average speed** is used. Average speed is the total distance covered divided by the time interval. Unless stated otherwise, speed is discussed relative to the surface of the Earth. Speed is a **scalar quantity** because it has amount or magnitude.

Velocity is a **vector quantity** because it includes amount or magnitude and direction. Velocity is the speed and direction of a moving object. Velocity is calculated the same way that speed is calculated and has the same units as speed but also includes direction. **Escape velocity** is the velocity that an object must reach in order to escape the gravitational influence of the Earth or other celestial body to which it is attracted. Escape velocity depends on the force of gravity for a planet or other celestial body. The more massive an object, the greater the force of gravity. The escape velocity on Earth is about 40,000 kilometers per hour. This means that a launched rocket must reach this velocity or it will fall back to Earth.

Acceleration describes how velocity changes over time. A change in velocity can be a change in speed or a change in direction. A force causes acceleration. The average acceleration is calculated by dividing the change in velocity by the change in time or:

$$\text{acceleration} = \frac{\text{final velocity} - \text{initial velocity}}{\text{final time} - \text{initial time}} \quad \text{or} \quad a = \frac{v_2 - v_1}{t_2 - t_1} = \frac{\Delta v}{\Delta t}$$

A negative acceleration is called deceleration. Examples of acceleration units include: miles per hour per second, kilometers per hour per second, meters per second per second (or meters per second squared), or feet per second per second (or feet per second squared).

Earth's gravity exerts a force on objects and can cause them to accelerate toward Earth. Objects that fall to the ground experience acceleration due to gravity. This acceleration is symbolized by **g**. For objects falling in a vacuum or neglecting air resistance, we will use a value of 9.8 m/s^2 or 32 ft/s^2 for g. Starting at rest, this means after one second a falling object has a velocity of 9.8 m/s or 32 ft/s, after two seconds 19.6 m/s or 64 ft/s, with the velocity, theoretically, continuing to increase. In fact, objects falling on Earth experience the downward force due to gravity and an upward force due to air resistance or the friction of the air. The force of friction increases with speed. When the amount of force due to friction is equal to the amount of force due to gravity, the object stops accelerating and moves with constant velocity called the **terminal velocity**.

Newton's First Law of Motion states: Every object continues in its state of rest, or of uniform motion in a straight line, unless it is made to change that state by forces acting upon it. Stated in a simplified way, this law says that whatever the present state of motion of an object, it will take force to change it. Sometimes this is called the law of inertia because **inertia** is the tendency of objects to resist changes in motion. A classic demonstration of inertia is when a tablecloth is quickly pulled from a tabletop leaving dishes that were on the tablecloth on the table.

Newton's Second Law of Motion states: The acceleration of an object is directly proportional to the net force acting on the object, is in the direction of the net force, and is inversely proportional to the mass of the object. With F=force, m=mass, and a=acceleration, this law can symbolized as: F = ma. This equation can be written in various ways to calculate force, mass, or acceleration.

Newton's Third Law of Motion states: Whenever one object exerts a force on a second object, the second object exerts an equal and opposite force on the first. Stated another way, the third law says for every action there is an opposite and equal reaction. A common example of the third law is the action force of a rocket pushing on gas that escapes from it with the reaction force being the gas pushing on the rocket and moving the rocket forward.

Energy

Matter and energy help make up the universe. Earlier, matter was identified as anything that has mass and takes up space (has volume). Frequently, energy is described as the ability to do work. Like work, energy can be measured in joules. Many forms of matter can be observed or detected using our senses. Energy exists in different forms such as light, heat, sound, and electricity. Einstein's famous equation, **E=mc^2**, where E stands for energy, m is mass, and c is the speed of light (in a vacuum, c is about 300,000 km/s or 186,000 miles/s) relates matter and energy.

Different sources classify forms of energy in slightly different ways. One way is to compare kinetic energy and potential energy (discussed in more detail later). Commonly listed other forms of energy include: **chemical** (due to composition; also considered a form of potential energy by some); **electrical** (due to the movement of charge or electrons; kinetic); **electromagnetic** (in the form of various kinds of rays or waves such as gamma rays, heat, light, microwaves, radio waves, and x-rays; kinetic); **magnetic energy** (due to magnetic attraction); **nuclear** (result from the splitting or joining of nuclei); and, **sound** (when objects or atoms vibrate; kinetic).

The **Law of Conservation of Energy** states that energy cannot be created nor destroyed; it may be transformed from one form to another but the total amount of energy never changes. This is true when systems are considered in their entirety. Thermodynamics involves the study of heat and represents a significant area of study in physics. The **First Law of Thermodynamics**, just like the Law of Conservation of Energy, states that within a system, energy is conserved as it changes from one form to another. Stated another way, the first law states that when energy is converted from one form to another, the same amount of energy exists after the conversion as before.

Fossil fuels include coal, oil, and natural gas. Fossil fuels are considered **non-renewable** sources of energy because once they are used, they are used up and unavailable for future use. Based on present use and estimated known reserves, projections are made as to how long fossil fuels resources will last. **Renewable** energy sources, such as solar energy and wind energy, are not viewed as having this limitation. What are advantages and disadvantages of renewable versus non-renewable energy sources? Among other things, most discussions attempting to answer this question would involve: availability/accessibility, pollution, cost, and how easy or difficult it is to convert the energy into a useful form. The search for alternative sources of energy seems to get particular public attention during times of high oil and gas prices.

Kinetic and Potential Energy

Potential energy (PE) is stored energy due to the position of a mass or due to the chemical makeup of a material. A rock resting on the edge of a cliff and a pendulum bob raised to one side both represent potential energy. A car battery has potential energy. Earlier, energy was described as the ability to do work. Thus, work units are used to measure energy. Potential energy due to position equals the work done to place it in position. Since $W = Fd$, potential energy equals force times distance applied ($PE = Fd$). The work done to lift an object is its weight (F) times the distance or height (h). From Newton's Second Law, $F = ma$, and since acceleration is the acceleration due to gravity, by substitution, $PE = mgh$.

Masses in motion are said to possess **kinetic energy**. Kinetic energy depends upon mass and velocity. Kinetic energy can be calculated using the formula: $KE = .5mv^2$ where m is the mass and v is the velocity. Note that if the mass unit is kilograms and the speed unit is m/s, the resulting unit for KE will be $\frac{kgm^2}{s^2}$ which is the same as a joule (J).
Recall, earlier joules were identified as a unit of work and energy.

Heat

Earlier Activity-Based Physical Science units provided some background information on heat. Review that information in addition to reading and studying the information that follows.

Heat is a form of energy associated with the movement of the atoms in a substance or in an object. The amount of heat energy depends on mass and temperature:

Heat absorbed or released = change in temperature x mass x specific heat capacity
$$H = \Delta t \times m \times C_\rho$$

Specific heat capacity is the ability of a material to absorb (or release) heat. It is measured in units of joules per gram degree Celsius (J/g °C) or calories per gram degree Celsius (cal/g °C). A joule is the amount of energy or work required to maintain a force of one newton through a distance of one meter. A calorie is the amount of energy needed to raise the temperature of one gram of water one Celsius degree. One calorie equals 4.186 joules.

The Second Law of Thermodynamics states that when heat energy is converted to mechanical energy some heat energy is wasted. Some of the heat energy given off by a source can be converted into useful work or mechanical energy but not all of the heat energy.

There is a **Zeroth Law of Thermodynamics**. Because it is a simple law, less time is spent on it than the first or second laws. This law states that heat flows from an object at higher temperature to the object at lower temperature. If there is no temperature difference, then the objects will not exchange heat.

Electricity

Electricity is the movement of charge. The charge can be the result of the movement of electrons, a collection of electrons, or ions. Earlier you learned that electrons have a negative charge. **Conductors** allow electrons to move through them easily. **Insulators** do not allow electrons to move through them easily. When an object gains electrons it has a negative charge. When an object loses electrons it has a positive charge. **Static electricity** is electric charge built up in one place. It is a charge at rest. An example of static electricity may be found when taking clothes out of a dryer rubbing a balloon on your hair, or walking across a carpeted floor.

Current is the movement of charge in a conductor. Current can be measured in **amperes** (sometimes shortened to amps). Electric current can be direct current, DC, or alternating current, AC. The flow of charges in one direction is direct current. A battery has two terminals, one with a negative sign and one with a positive sign. Electrons move from the repelling negative terminal toward the attracting positive terminal. Alternating current has electrons that move first in one direction and then in the opposite direction. Most residential and commercial circuits are AC because electric energy in the form of AC can easily be stepped up to a high voltage to be transmitted great distances with small heat losses. Then the electricity can be stepped down to convenient voltages where it is used.

Potential difference is synonymous with voltage difference and refers to the difference between the potentials of two points in an electric field. Potential difference is also called voltage. When two points having different electric potential (voltage) are connected with a conductor, charge flows from the point with greater potential (greater voltage) to the one with lower potential (lower voltage) as long as the potential difference exists. Potential difference is measured in volts. As an analogy, some students like to think of potential difference as the force that causes electrons to move. Electrons do not move with equal ease through all materials.

An electrical **circuit** is a closed path capable of being followed by electricity. A **series circuit** provides only one path for the electricity while a **parallel circuit** provides more than one path for the current. Typically, a **short circuit** is where a low-resistance connection is made accidentally between two points in an electric circuit. This causes the current to flow where it was not intended to flow. Conductors are connected in a complete circuit or **closed circuit** that allows current to flow but current will not flow in an **open circuit** because they are not connected in a way that allows flow.

Magnetism

Nearly everyone has encountered magnets in their life and we recognize that magnets come in a variety of shapes. Magnetism is believed to result from the movements of electrons within the atoms of substances. It is thought that in magnetic materials groups of atoms, called **domains**, all point in one direction and their magnetic forces add up. In nonmagnetic materials, the domains point in all directions and their domains tend to cancel out.

A **magnetic field** is the region in which a magnetic force can be detected. Every magnet is surrounded by an invisible magnetic field that becomes weaker as distance from the magnet increases. The magnetic field is strong closest to the magnet. The power of a magnet is strongest at the poles that are typically near the end of a bar-shaped magnet. Unlike poles attract and like poles repel. The poles of a magnet are named after the poles of the Earth to which they are attracted.

Electricity can be used to produce magnetism as in an electromagnet and magnetism can be used to produce electricity. When a coil of wire is moved through the force field of a magnet, electricity is produced. Many modern electric generators work on this principle.

Light

Scientists recognize two ways of understanding light and refer to it as the "wave-particle duality" of light. Light is connected to electrical and magnetic fields as might be inferred from the fact that light is a part of the electromagnetic spectrum. Light travels as waves so it has some features attributed to wave phenomena. In some circumstances, light seems to behave like tiny particles or packets of energy, called **photons**. In a vacuum, the **speed of light** is about 300,000 km/s, or 300,000,000 m/s, or 186,000 miles/s. For our purposes, it is sufficient to use these approximations for the speed of light when making calculations. Nothing has been found to travel faster than light.

Students use the memory device "**ROY G BIV**" to recall the colors of light that make up white light (red, orange, yellow, green, blue, indigo, violet). Red light has a longer wavelength than does violet light. The color of an object is determined by the color of the light reflected from its surface; other colors of light are absorbed. Colored filters operate in a similar way. A blue filter blocks all wavelengths of light except for blue.

Light can interact with materials in a variety of ways. Light can be transmitted through a material, reflected, refracted, absorbed, or scattered. Certain kinds of surfaces will reflect light. Many objects do not produce their own light but **reflect** light. Smooth, shiny surfaces produce clear, bright images because most of the light is reflected with very little light scattered. Mirrors are examples of such surfaces. For light, the angle of the incoming light is equal to the angle of the reflected light or **angle of incidence equals angle of reflection** (when measured from the normal, i. e., an imaginary line perpendicular to the point where the light strikes the object). Light bends or **refracts** when it changes speed while travelling through different substances. Light travels fastest in a vacuum. Lenses, prisms, and raindrops can refract light. The longer wavelengths of red light cause it to slow down the least and be refracted the least.

Laser stands for light amplification by the stimulated emission of radiation. The waves of laser light are of one wavelength. Feeding ordinary light or electricity into a medium such as ruby crystal produces this type of light. Atoms in the medium release light of a particular wavelength. Special mirrors build up the light energy and the light becomes intense enough to form a laser beam.

Sound

A **vibrating object** usually produces sound. A vibrating object in air pushes and pulls against the air molecules and causes some of the molecules be squeezed together and others to be stretched apart. The air molecules then squeeze and stretch the air molecules around them so that a wave of energy moves through the air. Sound can also travel through solids and liquids. Sound, unlike light, needs a medium through which to travel. The **pitch** of a sound depends on its **frequency**. The **Doppler effect** is an apparent change in pitch as a sound source approaches and then moves away. As the source of sound approaches, the pitch is higher, and as the source of sound moves away, the pitch is lower. Animals differ in their ability to hear sounds at different pitches. The **decibel scale** measures the intensity of the sound, which is similar to the loudness or volume of the sound. On the decibel (dB) scale, normal conversation is around 60 dB, around 10 dB is barely audible, and loud music is around 120 dB.

Nuclear Energy

Albert Einstein predicted a relationship between matter and energy and expressed it in his famous equation, $E=mc^2$, where E stands for energy, m is mass, and c is the speed of light (in a vacuum, c is about 300,000 km/s or 300,000,000 m/s or 186,000 miles/s). Thus, a small amount of mass can yield a large amount of energy. Two nuclear processes that release energy are **fission** and **fusion**. Some substances, like uranium, have unstable nuclei that when bombarded with neutrons or other atomic particles split apart releasing other neutrons and a burst of energy. Nuclear power plants make use of fission or the splitting apart of nuclei. Fusion involves the combination of two lightweight nuclei to form a heavier, more stable nucleus. In the Sun, hydrogen is the fuel that undergoes fusion. Scientists would like to develop nuclear power plants that use fusion to produce energy because hydrogen is readily available as a fuel (seawater would be a source) and because fusion would produce little or no radioactive waste. Thus far, problems like achieving the high temperatures necessary for fusion and containment have been insurmountable obstacles.

Both nuclear fission and nuclear fusion change matter (atoms) into energy. Scientists, working under certain conditions, have changed matter into energy.

The table below presents information about **three types of nuclear radiation**.

Table 6-6   Radiation

Type	Symbol	Charge	Mass	Can be stopped by
alpha particle	α	+2 ($2p^+$ and $2 n^0$)	4	paper
beta particle	β	-1 (e^-)	0	aluminum foil
gamma ray	γ	none	0	lead or concrete

Gamma radiation is the most penetrating and potentially the most dangerous. Beta decay occurs when a neutron inside a nucleus spontaneously decays into a proton and an electron. A proton can decay into a neutron and a positron. A positron is similar to an electron but with a positive charge. Positrons are also considered to be beta particles.

Scientists

Developments in physics have been closely connected to developments in mathematics, other sciences, and technology.

English scientist **Isaac Newton** used four simple laws to explain motions of objects on or near the Earth and motions of objects in space. The laws used were Newton's law of universal gravitation and his three laws of motion. This work was documented in 1687 when Newton published *Mathematical Principles of Natural Philosophy*. Newton invented a form of mathematics called calculus. As an aside, independently and at about the same time, mathematician Gottfried Leibniz developed calculus. Also, Newton made significant contributions to the study of light. Using prisms, he conducted experiments on light that led him to suggest white light was a mixture of colors. Newton suggested a particle theory of light that competed with Christiaan Huygens' wave theory of light.

Albert Einstein's profound insight changed human understanding of the world and resulted in products that changed the world. In 1905, he published papers on the Special Theory of Relativity, photoelectric effect, and the original version of $E=mc^2$. The relativity paper contributed to the use of atomic energy. Increased understanding of the photoelectric effect led to the development of vacuum tubes and integrated circuits that contributed to computers and their impact. In addition, his explanation of Brownian motion added to the understanding of the shape and size of molecules that eventually contributed to greater understanding of DNA.

Science Careers

Physicists describe the world around us by studying a range of phenomena from tiny particles of matter to large objects in the universe like galaxies. Physicists study matter and energy and their interactions. Physics contributes to all the sciences and serves as a foundation for much of technology. The work of physicists has contributed to many advances that have benefited society. Physics careers provide challenging opportunities in education, industry, and government. The work of physicists has a wide range of applications including contributions to computers, transportation, communication, medicine, safety, and the environment.

A college degree is required for physics careers and many jobs require an advanced degree. Contact the American Physical Society or the American Association of Physics Teachers for additional information about physics careers.

Physics Case Study – Laws for Heaven and Earth

As stated earlier, in 1687, Isaac Newton published *Mathematical Principles of Natural Philosophy*. In this work, Newton proposed one set of physics laws for objects at or near the Earth's surface and for heavenly bodies. Newton summarized and extended the work of such scientists as Galileo, Descartes, and Kepler. His analysis of gravitational force and motion showed planetary orbits had to be elliptical as Kepler proposed earlier. His three laws of motion describe relationships among force, mass, and acceleration. His law of universal gravitation states the force of gravity depends on the masses of the two objects and the distance between them. Newtonian physics explains many different phenomena and was accepted for centuries. Newton's ideas are still widely used even though Einstein's relativity extends beyond Newtonian explanations.

Physics Case Study – Matter, Energy, Time, and Space

Albert Einstein extended Newton's work by developing a more complex theory. Both Newton's and Einstein's works are great human accomplishments. Einstein formulated the special theory of relativity and later a general theory of relativity. One special relativity idea is that nothing can travel faster than the speed of light which is the same for all observers regardless of whether the observer or the light source is moving. Another special relativity idea is that any form of energy has mass and that matter is a form of energy. $E=mc^2$ where E symbolizes energy, m is mass, and c stands for the speed of light, illustrates the relationship between mass and energy. General relativity depicts Newton's gravitational force interacting with space. Many predictions based on the theory of relativity have been confirmed but scientists are still trying to confirm the existence of gravitational waves.

109 Physics Concepts and Terms List

absolute zero

acceleration

action/reaction

air resistance

alpha particle

amplitude

amperes (amps)

battery

beta particle

bioluminescence

calorie vs. Calorie

carbon-14

centripetal force

chain reaction

chemiluminescence

circuit (parallel, series)

circuit diagram (symbols)

color

conduction

conductor

conservation of energy

constellations

control rods

convection

coolant

crest

current (AC/DC)

decibel

direction

distance

Doppler effect

$E=mc^2$

efficiency

electricity

electromagnetic spectrum (gamma, x-ray, ultraviolet, visible, infrared, microwave, radio, television)

energy

energy alternatives

escape velocity

exposure (short, prolonged)

F=ma

fission

force (unbalanced, action, reaction)

fossil fuels

frequency

friction

fuel

fusion

gamma

gravity

half-life

heat

heat exchanger

heat transfer

horizontal velocity

horsepower

inertia

insulator

interference

ionization

joule

Kelvin

kinetic energy ($.5mv^2$)

laser

Law of Conservation of Energy

Laws of Thermodynamics

lens (concave, convex)

light

machine (simple, compound, lever, pulley, wheel and axle, inclined plane, screw, wedge)

magnetism

mass

matter

measurement

mechanical advantage

mirror

moderator

momentum (mv)

motion (relative)

motion in two directions

NASA/NASA Strategic Enterprises

Newton's Three Laws of Motion

non-renewable

nuclear reactor (components, 3 events, problems)

ohm

Ohm's Law

photon

pitch

polarized light

potential energy (mgh)

power

prism

radiation

radioactive decay

radon

reactor problems

reclamation

reflection

refraction

relative motion

renewable

resistance

safety shields

scalar

sound

specific heat

speed

system

temperature

terminal velocity

trough

vector

vibration

velocity

volume

volt

watt

wave (compressional, transverse)

wavelength

weight

weightlessness vs. microgravity

work (input, output)

Abraham, M. R. (1992). Instructional strategies designed to teach science concepts. In F. Lawrenz, K. Cochran, J. Krajcik, & P. Simpson (Eds.). Research matters . . . To the science teacher (pp. 41-50). Manhattan, KS: National Association of Research in Science Teaching.

American Association for the Advancement of Science. (1993). Benchmarks for science literacy. New York: Oxford University Press.

American Association for the Advancement of Science. (1990). Science for all Americans. New York: Oxford University Press.

American Geological Institute. (1965). Investigating the earth (laboratory manual pp.19-4 – 19-5 and teacher's guide pp. 19-21 – 19-24). Denver, CO: Smith-Brooks Printing Company (for laboratory manual) and Boulder, CO: Johnson Publishing Company (for teacher's guide).

Arons, A. (1977). The various language: An inquiry approach to the physical sciences. New York: Oxford University Press.

College Entrance Examination Board. (1986). Academic preparation in science: Teaching for transition from high school to college. New York, NY: College Board Publications.

Council for Educational Development and Research. (1993). Edtalk: What we know about science teaching and learning. Washington, DC: CEDR.

Duschl, R. A. (1990). Restructuring science education: The importance of theories and their development. New York: Teachers College Press.

Farmer, W. A., Farrell, M. A., & Lehman, J. R. (1991). Secondary science instruction: An integrated approach. Providence, RI: Janson Publications, Inc.

Harms, N., & Yager, R. (1981). Project synthesis. In What research says to the science teacher, (3) 9pp. 53-72).

Hazen, R. M., & Trefil, J. (1991). Science matters: Achieving scientific literacy. New York: Doubleday.

International Society for the Enhancement of Eyesight. Available January 4, 2002 at: http://www.i-see.org/ with specifications given for an eye chart available at a http://www.i-see.org/eyecharts.html.

Johnson, D., Johnson, R., Holubec, E. J., & Roy, P. (1986). <u>Circles of learning: Cooperation in the classroom</u>. Alexandria, VA: Association for Supervision and Curriculum Development.

Johnston, P., & Aber, S. W. (1990). <u>Introduction to earth science lab</u>. Edina, MN: Burgess International Group, Inc.

Kyle, W. C., Abell, S. K., & Shymansky, J. A. (1992). Conceptual change teaching and science learning. In F. Lawrenz, K. Cochran, J. Krajcik, & P. Simpson (Eds.). <u>Research matters . . . To the science teacher</u> (pp. 41-50). Manhattan, KS: National Association of Research in Science Teaching.

Matthews, M. R. (1994). <u>Science teaching: The role of history and philosophy of science</u>. New York: Routledge.

National Research Council. (2000). <u>How people learn: Brain, mind, experience, and school</u>. Washington, DC: National Academy Press.

National Research Council. (1996). <u>National science education standards</u>. Washington, DC: National Academy Press.

National Science Teachers Association. (1993). <u>The content core</u> (rev. ed.). Washington, DC: NSTA.

National Science Teacher Association. (1992). <u>Relevant research</u>. Washington, DC: Author.

National Science Teacher Association. (1992). <u>Scope, sequence, and coordination of secondary school science: Volume II relevant research</u>. Washington, DC: Author.

National Science Teacher Association. (1992). An NSTA position statement: Laboratory science. In <u>NSTA Handbook 1991-92</u> (pp. 166-169). Washington, DC: Author.

Novak, J. D., & Gowin, D. B. (1984). <u>Learning how to learn</u>. Cambridge: Cambridge University Press.

Tobias, S., & Tomizuka, C. T. (1992). <u>Breaking the science barrier: How to explore and understand the sciences</u>. New York: College Board Publications.